# EUROPEAN RAILWAYS MOTIVE POWER

## Volume 2 - Germany, Portugal & Spain

### 1st Edition

### Neil Webster

**metro**

ISBN 0-947773-52-5

# FOREWORD

This book is the second volume in a series which it is intended will be an enthusiast's guide to the motive power of the major railways of Europe. Each volume will have a strictly limited print-run so as to enable updated volumes to be issued approximately every two years. The size of print-run necessary to cost-effectively include colour photographs is such that this is not possible on our limited print-run. However, we feel sure our readers will appreciate our position - one of concentrating on producing an up-to-date series, rather than a pretty one. Watch out for our advertising in "Railway Magazine" and "European Report" for other volumes in the series. We apologise for being unable to advertise in the most appropriate magazine ("Today's Railways"), but the publisher of that magazine will not allow advertisements for what he considers to be "competing products"! Readers comments, observations and corrections are always welcome to amend, update and improve the series. These should be sent to the author via the publisher's address shown on the title page.

All comments will be gratefully acknowledged.

Neil Webster  August 1995.

# INDEX

# INTRODUCTION

Under each country heading, various useful information for the railway enthusiast is given, followed by listings of the motive power of both the national railway system and also the more major private railways and operators, but excluding industrial and tourist lines and preservation sites. Depot allocations are given where available/appropriate, with a listing of the codes used appearing under each individual railway heading. (Z) shown alongside a vehicle number or allocation indicates that the vehicle was stored out of service at the time of going to press.

For each vehicle type, brief technical details are given in metric units. Codes used to indicate the various builders/manufacturers/constructors are common throughout this volume and appear in a table at the end of this introduction. Abbreviations used for metric measures etc. are as follows:

| | | | |
|---|---|---|---|
| km | kilometres | m | metres |
| km/h | kilometres per hour | mm | millimetres |
| kW | kilowatts | rpm | revolutions per minute. |

For each country information regarding "Railrover" and similar tickets is given to assist the enthusiast in planning visits to Europe. All prices quoted are the latest available at the time of going to press and reflect the known position in August 1995 unless otherwise stated. Further details regarding tickets and fares may be obtained from the offices of the relevant railways or their agents at the addresses shown. Please do not direct any such enquiries at either the publishers or author.

# GUIDE TO BUILDERS ETC.

The following abbreviations are used for builders/equipment manufacturers in vehicle class headings:

| | |
|---|---|
| ABB | ASEA Brown Boveri Transportation AG. |
| AEG | Allgemeine Elekticitäts-Gesellschaft. |
| AFA | Akkumulatoren-Fabrik-Hagen. |
| Alco | Alco Products Inc. |
| Allan | N.V. Allan & Co. |
| Alsthom | Société Générale de Constructions Electriques et Méchaniques, Alsthom. |
| Atiensa | Aplicaciones Téchnicas Industriales SA. |
| B & L | SA des Établissements Brissoneau & Lotz. |
| Babcock | SEC Babcock & Wilcox CA. |
| Bautzen | VEB Waggonbau, Bautzen. |
| BBC | Brown, Boveri & Cie AG. |
| Bielhack | Martin Bielhack Maschinenfabrik und Hammerwerk GmbH. |
| Bombardier | Bombardier Inc. |
| Borsig | Borsig Lokomotivwerke GmbH. |
| Büssing | Büssing Automobilwerke AG. |
| CAF | Construcciones des Ateliers et Forges de la Loire. |
| Caterpillar | Caterpillar Inc. |
| Charkov | Lokomotivwerk Charkov. |
| Daimler-Benz | Daimler-Benz AG. |
| Deutz | Klöckner-Humboldt-Deutz AG. |
| Drewry | Drewry Car Co. Ltd. |
| Duewag | Duewag AG. |
| DWM | Duetsche Waggon- und Maschinenfabrik. |
| Esslingen | Maschinenfabrik Esslingen. |
| Fiat | Fiat SpA. |
| Fuchs | H. Fuchs Waggonfabrik AG. |
| Gardner | L. Gardner & Sons Ltd. |
| GEC | GEC Traction. |
| GEC Alsthom | GEC Alsthom. |
| GEE | General Electric Co. (Spain). |
| General Electric | General Electric Co. (USA). |
| Gmeinder | Gmeinder & Co. GmbH. |
| Görlitz | VEB Waggonbau, Görlitz. |
| Hartmann | Sächsische Maschinenfabrik Rich Hartmann AG. |
| Henschel | Rheinstahl Henschel AG. |
| Johannisthal | VEB Motorenwerk, Johannisthal. |
| Jung | Arnold Jung Lokomotivfabrik GmbH. |
| Kolomna | Lokomotivwerk Kolomna. |
| Krauss-Maffei | Krauss-Maffei AG. |
| Krupp | Friedrich Krupp Maschinenfabrik. |
| LEW | VEB Lokomotivbau und Elektrotechnische Werke, Hennigsdorf. |
| LHB | Linke Hoffman Busch. |
| LKM | VEB Lokomotivbau "Karl Marx", Babelsberg. |
| Lüttgens | Waggonfabrik Lüttgens GmbH. |
| Macosa | Material y Construcciones SA. |
| MaK | Mak Maschinenbau Kiel GmbH. |
| MAN | Maschinenfabrik Augsburg-Nürnberg AG. |
| Maybach | Maybach Motorenbau GmbH. |
| MBB | Messerschmitt-Bölchow-Blohm GmbH. |
| Melco | Mitsubishi Electric Corporation. |
| Meinfesa | Mediteranea de Industrias del Ferrocarril SA. |
| Mercedes | Daimler-Benz AG. |
| MLW | Montreal Locomotive Works. |
| Moyse | Établissements Gaston Moyse. |
| MTU | Motoren-und Turbinen Union GmbH. |
| Nohab | Nydqvist & Holm AB. |

3

| | |
|---|---|
| North British | North British Locomotive Co. Ltd. |
| O & K | Orenstein-Koppel und Lübecker Maschinenbau AG. |
| Oerlikon | Société Oerlikon |
| Orion | Orion Werke. |
| Rathgeber | Waggonfabrik Josef Rathgeber AG. |
| Rolls Royce | Rolls-Royce Ltd. |
| Roßlau | VEB Elbewerk, Roßlau |
| SACM | Société Alsacienne des Constructions Méchaniques. |
| Scania | Saab-Scania AB. |
| Scania Vabis | AB Scania-Vabis. |
| Schwartzkopff | Berliner Maschinenbau AG. |
| Siemens | Siemens AG. |
| SOREFAME | Sociedades Reunidas de Fabricaçôes Metálicas. |
| SSCM | Société Sugérienne de Constructions Méchanique. |
| Strömungsmaschinen | VEB Strömungsmaschinen. |
| Sulzer | Sulzer Brothers Ltd. |
| Uerdingen | Waggonfabrik Uerdingen. |
| Voith | J.M. Voith GmbH. |
| Voroshilovgrad | Voloshilovgrad Diesel Locomotive Works. |
| Vulcan | Schiffs und Maschinenbau Vulcan. |
| Waggon Union | Waggon Union GmbH. |
| Wegmann | Waggonfabrik Wegmann. |
| Windhoff | Rheiner Maschinenfabrik Windhoff. |
| Wismar | Triebwagen und Waggonfabrik Wismar. |
| WMD | Waggon und Maschinenbau GmbH. |
| 23 August | "23 August" Works. |

# FREEDOM PASS TICKETS (EURO DOMINO)

Euro Domino tickets are now the standard tourist ticket on all European railways. They offer either 3, 5 or 10 days unlimited rail travel within a pre-determined one month period on the national network of the country for which purchased in either First or Second Class, and are also valid on some private systems and ferries. Children usually travel at 50% of the adult fare, but age limits vary between countries. A Youth ticket (2nd Class only) is available for persons aged between the upper age limit for Child fares and 26 years of age.

Euro Domino tickets may be normally be purchased at any major station in any of the participating countries (all European countries except Albania, Romania and the former Soviet Republics), and depending on exchange rates it can be cheaper to the relevant country rather than in the UK - so you are advised to make enquiries regarding price and availability before travelling.

The ticket is dated by the user on each day used, and is valid from midnight to midnight on that day, except that for overnight trains travel is permitted after 19.00 on the day preceding the date marked. In Germany, travel is permitted from 23.00 on the day preceding, even on non-overnight services.

Leaflets advertising the tickets suggest no supplements are necessary on EC, IC or ICE services, although information elsewhere suggests that these may be payable on Swedish X2000 services.

Seat reservations, sleeping and couchette berths are not included in the ticket price.

Euro Domino purchasers in the UK are entitled to a reduction for travel from London to continental ports.

Where tickets other than Euro Domino are available, details are shown under the appropriate country heading. Many of these tickets offer better value than the Euro Domino, and travellers are advised to consider their requirements carefully before purchasing a ticket.

Prices for Euro Domino tickets in the countries covered by this volume are as follows:

| Country | Class | 3 Day Adult | 3 Day Youth | 5 Day Adult | 5 Day Youth | 10 Day Adult | 10 Day Youth |
|---------|-------|-------------|-------------|-------------|-------------|--------------|--------------|
| GERMANY | 1 | N/A | N/A | £67.00 | N/A | £125.00 | N/A |
| | 2 | £38.00 | £29.00 | £45.00 | £34.00 | £83.00 | £63.00 |
| PORTUGAL | 1 | £84.00 | N/A | £121.00 | N/A | £159.00 | N/A |
| | 2 | £55.00 | £39.00 | £79.00 | £59.00 | £106.00 | £79.00 |
| SPAIN | 1 | £93.00 | N/A | £135.00 | N/A | £173.00 | N/A |
| | 2 | £62.00 | £46.00 | £84.00 | £63.00 | £115.00 | £87.00 |

The above prices are as quoted by SNCF Railshops, London & Glasgow (Tel: 0345-300003). It should be noted prices quoted by SNCF Railshops are between £3.00 and £17.00 cheaper than the prices quoted by BR International for the same tickets!!

Proof of age and passport number details required as appropriate.

# INTER-RAIL TICKETS

The following tickets are available from BR stations with International booking facilities, many Travel Agents and British Rail International, International Rail Centre, Victoria Station, London, SW1V 1JY (Tel: 0171-834-2345, FAX 0171-922-9874).

### Inter-Rail Pass

Available to persons aged under 26 years only, who have been resident in Britain or Northern Ireland for at least six months and hold a valid passport. This ticket is divided into zones as follows:

A       Republic of Ireland
B       Norway, Sweden, Finland
C       Germany, Switzerland, Austria, Denmark
D       Czech Republic, Slovakia, Poland, Hungary, Bulgaria, Romania, Croatia
E       France, Belgium, Netherlands, Luxembourg
F       Spain, Portugal, Morocco
G       Italy, Greece, Turkey, Slovenia, ADN/HML ferry service between Brindisi (Italy) and
        Patras (Greece).

Valid for travel 2nd Class only. EC, IC, ICE, X2000, AVE and all other supplements must be paid at the time of useage.

A one zone ticket is valid for 15 consecutive days, with tickets for two or more zones valid for one month (consecutive days).

A reduction is granted to Pass holders for rail travel in Britain and Northern Ireland, as well as the rail portion of tickets between London and

the Continental Ports.

In addition, reduced fares are granted to Pass holders for services between Britain and Europe operated by Stena Sealink Line, Hoverspeed,

Oostende Lines, P & O European Ferries, Olau Line, Scandinavian Seaways, B & I Line and Irish Ferries.

Reduced prices are also applicable to Pass holders on other shipping services (e.g. in the Mediterranean and around Scandinavia), private railways in Switzerland and a number of other European countries, and free admittance is given to some transport museums.

Prices:   1 Zone       £179.00
          2 Zones      £209.00
          3 Zones      £229.00
          All Zones    £249.00

Passport required at the time of booking.

# GERMANY

**Inter-Rail 26+ Pass**
Available to persons who have been resident in Britain or Northern Ireland for at least six months and hold a valid passport.
Valid for 15 days or one month's unlimited 2nd Class travel (consecutive days) on the national rail systems of Austria, Bulgaria, Croatia, Czech Republic, Denmark, Finland, Germany, Greece, Hungary, Luxembourg, Netherlands, Norway, Poland, Romania, Republic of Ireland, Slovakia, Slovenia, Sweden and Turkey, plus ADN/HML Ferry Services between Brindisi (Italy) and Patras (Greece).
Valid for travel 2nd Class only. EC, IC, ICE, X2000, AVE and all other supplements must be paid at the time of useage.
Discounts are available for Pass holders on services between London and Hoek van Holland using the Stena Sealink line ferries via Harwich. Olau Lines, Scandinavian Seaways and a number of other shipping companies also offer discounts to Pass holders. Discounted fares are available to Pass holders for travel through Belgium to the frontiers with Germany, Luxembourg and Netherlands.
Prices:  15 Days     £209.00
         1 Month      £269.00
Passport required at the time of booking.

# GERMANY

## VISAS

Not required by UK citizens (for stays of up to 2 months). Full UK Passport necessary. British Visitors Passport not valid after 31.12.95. .

## BRITISH EMBASSY

**Address:** Friedrich-Ebert-Allee 77, W-5300 Bonn 1.(Tel: (228) 23 40 61; FAX: (228) 23 40 70 or 23 40 58).
There are British Consulates in Berlin, Bremen, Düsseldorf, Frankfurt am Main, Freiburg (Breisgau), Hamburg, Hanover, Kiel, München, Nürnberg and Stuttgart.

## CURRENCY & BANKING

Deutsche Mark (DM). 1 Mark = 100 Pfennigs. Notes are in denominations of DM 1,000, 500, 200, 100, 50, 20, 10 and 5. Coins are in denominations of DM 5, 2 and 1, and 50, 10, 5, 2 and 1 Pfennigs.
Banking hours are usually 0830-1300 and 1430-1600 Monday to Friday only, with opening until 1730 or 1800 on Thursdays.
Visa, Access/Mastercard and American Express cards are widely accepted. Travellers Cheques are readily exchanged, and Eurocheques may be cashed at many banks.

## OTHER USEFUL INFORMATION

**Language:** German.
**Time:** GMT + 1 hour (GMT + 2 hours in summer).
**Electricity:** 220 V ac, 50 Hz. Continental 2-pin plugs are standard.
**Usual Public Holidays:** New Year's Day, Epiphany (Jan 6), Good Friday, Easter Monday, May Day, Ascension Day, Whit Monday, Corpus Christi, Assumption, Unity Day (Oct 3), All Saints' Day, Repentance Day (Nov 22), Christmas Day, December 26.
Full details of these and other local holidays may be obtained from the German National Tourist Office.

## TOURIST INFORMATION

German National Tourist Office, Nightingale House, 65 Curzon Street, London, W1Y 7PE. (Tel: 0171-495-3990; FAX 0171-495-6129).

# Deutsche Bahn AG (DBAG).

**Gauges:** 1435 mm; 900 mm; 750 mm; 600 mm.
**Route Length:** 40466 km (1435 mm), 267 km (narrow gauges).
**Standard Electrification System:** 15 kV 16.67 Hz ac overhead.
**Other Electrification Systems:** 25kV 50Hz overhead; 500V dc overhead; 1200V dc side contact third rail

# NUMBERING

Each vehicle is allocated a six digit number, followed by a computer check digit. The first three digits of the number denote the class, and the second three digits the serial number of the vehicle. The first digit of the number denotes the type of vehicle as follows:

| | |
|---|---|
| 0 | Steam Locomotive. |
| 1 | Electric Main Line Locomotive. |
| 2 | Diesel Main Line Locomotive. |
| 3 | Shunting Locomotive/Tractor. |
| 4 | Electric Multiple Unit. |
| 5 | Battery Electric Multiple Unit. |
| 6 | Diesel Multiple Unit. |
| 7 | Departmental Vehicle or Diesel Railcar. |
| 8 | Electric Multiple Unit trailer. |
| 9 | Diesel Multiple Unit or Railcar trailer. |

The second and third digits of the number are used to differentiate between the different types of vehicle within a type of traction. An exception to this is with regard to narrow gauge vehicles, where the second and third digits of the number are always 99, irrespective of the vehicle type.
The fourth, fifth and sixth digits of a standard gauge vehicle and the fifth and sixth digits of a narrow gauge vehicle denote the individual vehicle serial number. For narrow gauge vehicles the fourth digit indicates the gauge of the vehicle as follows:

| | |
|---|---|
| 1 | 1000 mm gauge. |
| 6 | 600 mm gauge. |
| 7 | 750 mm gauge. |
| 9 | 900 mm gauge. |

In the case of multiple unit trailer vehicles, non driving trailers are always allocated serial numbers between 001 and 599, and driving trailers between 601 and 999.
The seventh digit is a computer check digit used to verify that an entry made to the railway computer is correct. These digits are actually carried on the vehicles, but are not shown in this book. The computer check digit may be calculated for any vehicle as follows:

| (155 236) | 1 | 5 | 5 | 2 | 3 | 6 |
|---|---|---|---|---|---|---|
| Multiply by | 1 | 2 | 1 | 2 | 1 | 2 |
| Answer | 1 | 10 | 5 | 4 | 3 | 12 |

Sum of digits of answer = 1+1+0+5+4+3+1+2 = 17
Subtract from next whole multiple of 10, viz 20-17 = 3
Check digit for 155 236 is thus 3.

# DEPOTS

The following depots have an allocation and are shown together with the codes used to denote them in this book.

**HAMBURG DIVISION**

| | |
|---|---|
| AFL | Flensburg |
| AH1 | Hamburg Eidelstedt |
| AH4 | Hamburg Wilhelmsburg |
| AK | Kiel |
| AL | Lübeck |
| AOP | Hamburg Ohlsdorf |

**BERLIN DIVISION**

| | |
|---|---|
| BCS | Cottbus |
| BFG | Frankfurt (Oder) |
| BHF | Berlin Hbf |
| BHW | Hoyerswerda |
| BPKR | Berlin Pankow |
| BWUR | Wustermark |

# GERMANY

## DRESDEN DIVISION
| | |
|---|---|
| DC | Chemnitz |
| DG | Görlitz |
| DH | Dresden Hbf |
| DR | Riesa |
| DRC | Reichenbach (Vogtl) |
| DZW | Zwickau (Sachs) |

## ESSEN DIVISION
| | |
|---|---|
| EDO | Dortmund Bbf |
| EE | Essen |
| EHG | Hagen Eckesey |
| EHM | Hamm (Westfalia) |
| EOB | Oberhausen Osterfeld-Süd |
| ESIE | Siegen |
| EWAN | Wanne-Eickel |

## FRANKFURT (MAIN) DIVISION
| | |
|---|---|
| FD | Darmstadt |
| FF1 | Frankfurt (Main) Hbf |
| FF2 | Frankfurt (Main) Rbf |
| FFG | Frankfurt (Main) Griesheim |
| FG | Gießen |
| FK | Kassel |

## HANNOVER DIVISION
| | |
|---|---|
| HB | Bremen Hbf |
| HBH | Bremerhaven |
| HBS | Braunschweig |
| HE | Emden |
| HG | Göttingen |
| HH | Hannover Hbf |
| HO | Osnabrück |
| HOLD | Oldenburg (Old) |
| HS | Seelze |
| HWG | Wangerooge |

## KÖLN DIVISION
| | |
|---|---|
| KA | Aachen West |
| KD | Düsseldorf Hbf |
| KG | Gremberg |
| KK | Köln Deutzerfeld |
| KM | Mönchengladbach |

## HALLE DIVISION
| | |
|---|---|
| LG | Güsten |
| LH1 | Halle G |
| LH2 | Halle P |
| LHB | Halberstadt |
| LL1 | Leipzig Süd |
| LL2 | Leipzig West |
| LLW | Leipzig Wahren |
| LM | Magdeburg |
| LMR | Magdeburg-Rothensee |
| LS | Stendal |
| LW | Lutherstadt-Wittenberg |

## MÜNCHEN DIVISION
| | |
|---|---|
| MA | Augsburg |
| MFL | Freilassing |
| MH1 | München Hbf |
| MH6 | München Steinhausen |
| MIH | Ingolstadt |
| MKP | Kempten (Allgäu) |
| MMF | Mühldorf (Oberbay) |

## NÜRNBERG DIVISION
| | |
|---|---|
| NHO | Hof |
| NLF | Lichtenfels |
| NN1 | Nürnberg Hbf |
| NN2 | Nürnberg Rbf |
| NRH | Regensburg |
| NWH | Würzberg |

## KARLSRUHE DIVISION
| | |
|---|---|
| RF | Freiburg (Breisgau) |
| RHL | Haltingen |
| RK | Karlsruhe |
| RM | Mannheim |
| RO | Offenburg |
| RSI | Singen (Hohentweil) |

## SAARBRÜCKEN DIVISION
| | |
|---|---|
| SKL | Kaiserslautern |
| SSH | Saarbrücken Hbf |
| STR | Trier |

## STUTTGART DIVISION
| | |
|---|---|
| TH | Heilbronn |
| TK | Kornwestheim |
| TP | Plochingen |
| TS | Stuttgart |
| TT | Tübingen |
| TU | Ulm |

## ERFURT DIVISION
| | |
|---|---|
| UE | Erfurt |
| UEI | Eisenach |
| UG | Gera |
| UM | Meiningen |
| UN | Nordhausen |
| US | Saalfeld (Saale) |
| UW | Weißenfels |

## SCHWERIN DIVISION
| | |
|---|---|
| WNT | Neustrelitz |
| WP | Pasewalk |
| WR | Rostock |
| WS | Schwerin |
| WST | Stralsund |
| WW | Wittenberge |

In addition, two workshops have allocations:
| | |
|---|---|
| HHX | Hannover-Leinhausen |
| KOPLX | Opladen |

# OTHER DEPOTS & STABLING POINTS

There are a considerable number of other depots and stabling points, as can be seen from the following list:

**HAMBURG DIVISION**
Altmannbrücke, Buchholz (Nordheide), Büchen, Cuxhaven, Hamburg Altona, Hamburg Harburg, Hamburg Hbf, Husum, Itzehoe, Lüneburg, Maschen, Niebüll, Neumünster, Pinneberg, Puttgarden, Rendsburg, Stade, Westerland (Sylt).

**BERLIN DIVISION**
Basdorf, Berlin Grunewald, Berlin Lichtenberg, Berlin Rummelsburg, Berlin Schöneweide, Berlin Wuhlheide Rbf, Elsterwerda, Eisenhüttenstadt, Frankfurt (Oder) Pbf, Fürstenwalde (Spreewalde), Guben, Jüterbog, Kietz, Königs Wusterhausen, Luckau, Lübbenau (Spreewald), Oranienburg, Seddin, Senftenberg, Wriezen.

**DRESDEN DIVISION**
Adorf (Vogtl), Altenberg (Erzgebirge), Amsdorf (bei Dresden), Annaberg-Buchholz, Aue (Sachs), Bad Schandau, Bautzen, Bischofswerda, Chemnitz Hilbersdorf, Döbeln, Dresden Zwickauer Straße, Falkenstein (Vogtl), Freiberg (Sachs), Freital, Glauchau (Sachs), Großenhain, Kamenz (Sachs), Königsbrücke, Kurort Oberwiesenthal, Löbau (Sachs), Neustadt (Sachs), Nossen, Oelsnitz (Erzgebirge), Plauen (Vogtl), Pima, Pockau-Lengefeld, Radebeul, Rochlitz (Sachs), Stollberg (Sachs), Werdau, Zittau.

**ESSEN DIVISION**
Altenhundem, Bestwig, Betzdorf (Sieg), Bielefeld, Bochum Langendreer, Borken (Westfalia), Coesfeld (Westfalia), Dorsten, Dortmund Gbf, Duisburg Hbf, Duisburg-Ruhrort Hafen, Duisburg-Wedau, Erndtebrück, Emmerich, Finnentrop, Gronau (Westfalia), Hagen Vorhalle, Kreuztal, Lage (Lippe), Lünen Süd, Münster (Westfalia), Neheim-Hüsten, Neubeckum, Oberhausen Hbf, Oberhausen West, Paderborn, Recklinghausen Ost, Schwerte (Ruhr), Soest, Unna, Warendorf, Wesel.

**FRANKFURT (MAIN) DIVISION**
Alzey, Bad Homburg, Bebra, Bingerbrück, Darmstadt-Kranichstein, Dillenburg, Erbach (Odenwald), Frankenberg (Eder), Frankfurt (Main) Süd, Friedberg (Hess), Fulda, Gladenbach, Goddelau-Erfelden, Hanau, Hungen, Korbach, Limburg (Lahn), Mainz, Mainz-Bischofsheim, Marburg, Nidda, Niedernhausen (Taunus), Ober Roden, Offenbach (Main), Rüdesheim (Rhein), St. Goarshausen, Stockheim (Oberhess), Treysa, Volkmarsen, Wabern, Weilburg, Werburg, Wiebelsbach-Heubach, Wiesbaden, Worms.

**HANNOVER DIVISION**
Altenbeken, Bremen-Grolland, Bremen-Inlandshafen, Bremen Rbf, Bremen Zollausschluß, Bremerhaven Seehafen, Bremerhaven-Speckenbüttel, Goslar, Hameln, Hannover Hgbf, Hannover-Linden, Helmstedt, Hildesheim, Holzminden, Kreiensen, Leer (Ostfr), Lehrte, Löhne, Minden (Westfalia), Neinberg (Weser), Nordenham, Northeim (Han), Ottbergen, Rhaden (Kr Lubecke), Rheine, Soltau (Han), Uelzen, Warburg (Westfalia), Wilhelmshaven.

**KÖLN DIVISION**
Aachen Hbf, Altenkirchen (Westerwald), Andernach, Au (Sieg), Dieringhausen, Düren, Düsseldorf-Derendorf, Euskirchen, Herzogenrath, Koblenz, Koblenz-Lützel, Köln Bbf, Köln Hbf, Köln-Eifeltor, Köln-Kalk Nord, Köln Nippes, Kreuzberg (Ahr), Jünkerath, Kaldenkirchen, Kleve, Krefeld, Mayen, Neuss, Neuweid, Oberlahnstein, Opladen, Remagen, Siershahn, Stolberg (Rheinland), Remscheid-Lennep, Troisdorf, Wuppertal, Xanten.

**HALLE DIVISION**
Altenburg, Aschersleben, Blankenberg (Harz), Bitterfeld, Borna (bei Leipzig), Brandenburg, Burg (bei Magdeburg), Dessau, Eilenburg, Eisleben (bei Magdeburg), Engelsdorf, Falkenberg (Elster), Gerbstedt, Haldensleben, Jerichow, Köthen, Leipzig-Plagwitz, Leipzig-Stötteritz, Merseburg, Oebisfelde, Querfurt, Rathenow, Röblingen am See, Roßlau, Salzwedel, Torgau, Wurzen, Zeitz.

**MÜNCHEN DIVISION**
Aschau (Chiemgau), Buchloe, Donauwörth, Eichstatt, Garmisch-Partenkirchen, Krumbach (Schwab), Landshut (Bay), Lindau, Memmingen, München Nord, Murnau, Neustadt (Donau), Nördlingen, Rosenheim, Schongau, Traunstein, Treuchtlingen, Wasserburg (Inn), Weiheim (Oberbay).

**NÜRNBERG DIVISION**
Amberg, Ansbach, Aschaffenburg, Bamberg, Bayreuth, Coburg, Gemünden (Main), Kirchenlaibach, Passau, Plattling, Pressig-Rothenkirchen, Marktredwitz, Schwandorf, Schweinfurt, Straubing, Weiden (Oberpf).

**KARLSRUHE DIVISION**
Bruchsal, Germersheim, Gründstadt, Heidelberg, Landau (Pfalz), Ludwigshafen (Rhein), Neckarelz, Neustadt (Weinstr), Neustadt (Schwarzwald), Rastatt, Villingen (Schwarzwald), Weinheim (Bergstraße).

**SAARBüCKEN DIVISION**
Cochem (Mosel), Dillingen (Saar), Einsiedlehof, Gerolstein, Homburg (Saar), Kirn, Kusel, Lauterecken-Grumbach, Lebach, Neukirchen (Saar), Pirmasens, Simmern, St. Wendel, Turkismühle.

**STUTTGART DIVISION**
Aalen, Aulendorf, Crailsheim, Freudenstadt, Friedrichshafen, Geislingen (Steige), Horb, Lauda, Pforzheim, Rottweil.

**ERFURT DIVISION**
Arnstadt, Artern, Eisfeld, Göscwitz (Saale), Gotha, Greiz, Ilmenau, Leinefelde, Lobenstein, Mühlhausen (Thür), Naumberg (Saale), Probstzella, Sangerhausen, Schleusingen, Schmalkalden, Sonneberg (Thür), Suhl, Triptis, Vacha, Weimar.

# GERMANY

## SCHWERIN DIVISION

Angermünde, Anklam, Bad Kleinen, Barth, Demmin, Eberswalde, Gnoien, Greifswald, Güstrow, Hagenow Land, Kühlingsborn Ostseebad, Löwenberg (Mark), Ludwigslust, Malchin, Meyenburg, Mukran, Neubrandenburg, Neuruppin, Neustadt (Dosse), Parchim, Prenzlau, Putbus, Rheinsberg (Mark), Rostock Seehafen, Saßnitz, Seebad Heringsdorf, Tempin, Waren (Müritz), Warnemünde, Wismar, Wittstock (Dosse).

# WORKSHOPS

**Bremen Seebaldsbrück:** Diesel Locomotives of Classes 211-218 and 290 & 291.
**Chemnitz:** Diesel Locomotives of Classes 219, 228, 229, 344-346 and various other Shunting Locomotives/Tractors.
**Cottbus:** Diesel Locomotives of Classes 232, 234 & 242.
**Dessau:** Electric Locomotives of Classes 112, 142, 143, 150, 155, 156, 171, 180 and also collision damage repairs to other Electric Locomotives.
**Görlitz:** Narrow Gauge Steam Locomotives.
**Halle:** Shunting Locomotives and Tractors, mainly ex DR types.
**Kassel:** Diesel Multiple Units, Railcars and Shunting Locomotives of Classes 360-365.
**Limburg:** Battery Electric Railcars.
**Meiningen:** Standard Gauge Steam Locomotives.
**München Freimann:** Electric Locomotives of Classes 110, 111, 139, 140, 151.
**Nürnberg:** Electric Multiple Units.
**Opladen:** Electric Locomotives of Classes 103, 110, 111, 113, 120, 140, 151, 181 and 184.
**Stendal:** Diesel Locomotives of Classes 201, 202, 204, 293 & 298.
**Wittenberge:** Diesel Multiple Units.

# RAILROVER INFORMATION

## Regional Passes

5 or 10 days unlimited travel (not necessarily consecutive) within a pre-designated 21 day period on all DBAG trains (including all supplements except on ICE services, seat reservations, couchette or sleeper berths) within the selected area. Regional Passes are not valid on bus services. Fifteen different area Passes are available as follows:

|  | |
|---|---|
| Region 101: | Hamburg/Schleswig Holstein. |
| Region 102: | Rostock/Schwerin/Baltic Coast. |
| Region 103: | Bremen/Münster/North Sea Coast. |
| Region 104: | Hannover/Lower Saxony. |
| Region 105: | Hannover/Berlin/Harz. |
| Region 106: | Berlin/Brandenburg. |
| Region 107: | Köln/Düsseldorf/Rhine-Ruhr. |
| Region 108: | Frankfurt/Hesse. |
| Region 109: | Erfurt/Weimar/Thuringen. |
| Region 110: | Dresden/Leipzig/Sachsen. |
| Region 111: | Köln/Mainz/Rhine-Moselle. |
| Region 112: | Stuttgart/Neckar Schwabia/Schwarzwald/Bodensee. |
| Region 113: | Nürnberg/Franken/Bayerisch Wald. |
| Region 114: | München/Bavarian Alps. |
| Region 115: | Berlin/Dresden/Leipzig. |

Regional Passes are available in three forms and are all the same price irrespective of area as follows:

Single Pass (Adult fares)
5 Days  - 1st Class £85.00; 2nd Class £54.00
10 Days - 1st Class £128.00; 2nd Class £85.00

Twin Pass (Any two adults travelling together)
5 Days  - 1st Class £128.00; 2nd Class £85.00
10 Days - 1st Class £192.00; 2nd Class £128.00

Family Pass (One or two parents or grandparents with any number of children/grandchildren under 16 years of age)
5 Days  - 1st Class £153.00; 2nd Class £102.00
10 Days - 1st Class £230.00; 2nd Class £153.00

Proof of identity (i.e passport) is required at the time of purchase of any of the above tickets and must also be carried whilst travelling.

For details of Freedom Pass (Euro Domino) tickets, please see the section at the front of this book.

**Tickets & Details**
Further details and literature on all the above passes and rail travel in general in Germany is available from German Rail, Suite 4, The Sanctuary, 23 Oakhill Grove, Surbiton, Surrey, KT6 6DU.

## TIMETABLE

The DB "Kursbuch" is published twice a year and is available from main DB stations and at station bookshops. It is also available in the UK from European Rail Timetables (Tel: 01909-485855), 39 Kilton Glade, Worksop, Nottinghamshire, S81 0PX.

## 1435 mm GAUGE STEAM LOCOMOTIVES

**Note:** Numbers shown are those actually carried. Where these differ from allocated computer numbers, these latter are shown in parentheses.

### CLASS 044                                    2-10-0

**Built:** 1926-44 by Wiener Lokomotivfabrik AG, Floridsdorf, Wien, Austria.
**Boiler Pressure:** 16 kg/cm$^2$     **Cylinder Diameter (3):** 550 x 660 mm.
**Length Overall:** 22.620 m. (Loco 13.817m; Tender 8.803 m.)
**Driving Wheel Diameter:** 1400 mm.
**Weight:** 174.20 tonnes.     **Maximum Speed:** 80 km/h.

44 2115   (044 115)  (Z)

### CLASS 050                                    2-10-0

**Built:** 1938-43 by Henschel/Kraus Maffei. Rebuilt 1958-63.
**Boiler Pressure:** 16 kg/cm$^2$     **Cylinder Diameter (2):** 600 x 660 mm.
**Length Overall:** 22.940 m. (Loco 13.680 m; Tender 9.260 m.)
**Driving Wheel Diameter:** 1400 mm.
**Weight:** 146.40 tonnes.     **Maximum Speed:** 80 km/h.

50 3521   (050 521)  WP       50 3565   (050 565)  (Z)       50 3707   (050 707)  (Z)
50 3522   (050 522)  (Z)

### CLASS 050                                    2-10-0

**Built:** 1943-48 by various builders. Rebuilt 1965-67 by DR, Stendal Works.
**Boiler Pressure:** 16 kg/cm$^2$     **Cylinder Diameter (2):** 600 x 660 mm.
**Length Overall:** 22.940 m. (Loco 13.680 m; Tender 9.260 m.)
**Driving Wheel Diameter:** 1400 mm.
**Weight:** 129.80 tonnes.     **Maximum Speed:** 80 km/h.

52 8017   (052 017)  (Z)       52 8075   (052 075)  (Z)       52 8097   (052 097)  (Z)
52 8021   (052 021)  (Z)       52 8079   (052 079)  (Z)       52 8134   (052 134)  (Z)
52 8063   (052 063)  (Z)       52 8087   (052 087)  (Z)       52 8145   (052 145)  (Z)

## 750 mm GAUGE STEAM LOCOMOTIVES

### 099 701-713                                  0-4-4-0T

**Built:** 1899-1922 by Hartmann. Rebuilt 1962 by DR.
**Boiler Pressure:** 15 kg/cm$^2$     **Cylinder Diameter (4):** (2) 240 x 380 mm,
                                                            (2) 370 x 380 mm.
**Length:** 9.000 m.     **Driving Wheel Diameter:** 760 mm.
**Weight:** 26.80-29.30 tonnes.     **Maximum Speed:** 30 km/h.

099 701 DR        099 705 DR        099 711  (Z)        099 712  (Z)        099 713   DR

# GERMANY

## 099 720      0-10-0T

**Built:** 1927 by Hartmann.
**Boiler Pressure:** 14 kg/cm²      **Cylinder Diameter (2):** 430 x 400 mm.
**Length:** 8.990 m.      **Driving Wheel Diameter:** 800 mm.
**Weight:** 42.25 tonnes.      **Maximum Speed:** 30 km/h.

099 720 DR

## 099 722-735      2-10-2T

**Built:** 1928-33 by Hartmann/Schwartzkopff.
**Boiler Pressure:** 14 kg/cm²      **Cylinder Diameter (2):** 450 x 400 mm.
**Length:** 10.450 m.      **Driving Wheel Diameter:** 800 mm.
**Weight:** 56.70 tonnes.      **Maximum Speed:** 30 km/h.

| | | | | |
|---|---|---|---|---|
| 099 722 (Z) | 099 725 DR | 099 728 DG | 099 731 DG | 099 734 DR |
| 099 723 DR | 099 726 DR | 099 729 DG | 099 732 DC | 099 735 (Z) |
| 099 724 DG | 099 727 DR | 099 730 (Z) | 099 733 DG | |

## 099 736-757      2-10-2T

**Built:** 1952-56 by LKM.
**Boiler Pressure:** 14 kg/cm²      **Cylinder Diameter (2):** 450 x 400 mm.
**Length:** 11.300 m.      **Driving Wheel Diameter:** 800 mm.
**Weight:** 55.00 tonnes.      **Maximum Speed:** 30 km/h.

| | | | | |
|---|---|---|---|---|
| 099 736 DR | 099 741 DR | 099 746 WSR | 099 750 DC | 099 754 (Z) |
| 099 737 DC | 099 742 DR | 099 747 DR | 099 751 DG | 099 755 (Z) |
| 099 738 DC | 099 743 DR | 099 748 WSR | 099 752 DR | 099 756 DR |
| 099 739 DR | 099 744 (Z) | 099 749 DC | 099 753 DR | 099 757 DC |
| 099 740 (Z) | | | | |

## 99 770-771      0-8-0T

**Built:** 1914/25 by Vulcan.
**Boiler Pressure:** 12 kg/cm²      **Cylinder Diameter (2):** 350 x 400 mm.
**Length:** 8.000 m.      **Driving Wheel Diameter:** 850 mm.
**Weight:** 24.00 tonnes.      **Maximum Speed:** 30 km/h.

099 770 WSR      099 771 WSR

## 099 780-781      0-8-0T

**Built:** 1938 by Henschel.
**Boiler Pressure:** 13 kg/cm²      **Cylinder Diameter (3):** 360 x 410 mm.
**Length:** 9.440 m.      **Driving Wheel Diameter:** 850 mm.
**Weight:** 32.40 tonnes.      **Maximum Speed:** 40 km/h.

099 780 WSR      099 781 WSR

## 099 782

**Built:**
**Boiler Pressure:**      **Cylinder Diameter:**
**Length:**      **Driving Wheel Diameter:**
**Weight:**      **Maximum Speed:**

099 782 WSR

# 900 mm GAUGE STEAM LOCOMOTIVES

## 099 901-903      2-8-2T

**Built:** 1932 by O & K.
**Boiler Pressure:** 14 kg/cm²      **Cylinder Diameter (2):** 380 x 550 mm.
**Length:** 10.595 m.      **Driving Wheel Diameter:** 1100 mm.

**Weight:** 43.68 tonnes.  **Maximum Speed:** 50 km/h.

099 901  WR    099 902  WR    099 903  WR

## 099 904-905    0-8-0T

**Built:** 1950-51 by LKM. Acquired by DR in 1961 from Wismut AG.
**Boiler Pressure:** 14 kg/cm².  **Cylinder Diameter (2):** 370 x 400 mm.
**Length:** 8.860 m.  **Driving Wheel Diameter:** 800 mm.
**Weight:** 32.40 tonnes.  **Maximum Speed:** 35 km/h.

099 904  WR    099 905  WR

# ELECTRIC LOCOMOTIVES

## CLASS 103    Co-Co

**Built:** 1970-74 by Henschel/Krauss-Maffei/Krupp.
**Electrical Equipment:** Siemens/AEG/BBC.  **Continuous Rating:** 7440 kW.
**Weight:** 114.00 tonnes.  **Maximum Speed:** 200 km/h.
**Length:** 19.500 m. (* 20.200 m.)  **Train Supply:** Electric.

| | | | | | | | | | |
|---|---|---|---|---|---|---|---|---|---|
| 103 101 | FF1 | 103 131 | FF1 | 103 160 | FF1 | 103 188 | AH1 | 103 218* | AH1 |
| 103 102 | FF1 | 103 132 | FF1 | 103 161 | FF1 | 103 189 | FF1 | 103 219* | AH1 |
| 103 103 | FF1 | 103 133 | FF1 | 103 162 | FF1 | 103 190 | AH1 | 103 220* | AH1 |
| 103 104 | FF1 | 103 135 | FF1 | 103 163 | FF1 | 103 191 | AH1 | 103 221* | AH1 |
| 103 105 | FF1 | 103 136 | FF1 | 103 164 | FF1 | 103 192 | AH1 | 103 223* | AH1 |
| 103 107 | FF1 | 103 137 | FF1 | 103 165 | FF1 | 103 193 | AH1 | 103 224* | AH1 |
| 103 108 | FF1 | 103 138 | FF1 | 103 166 | FF1 | 103 194 | AH1 | 103 225* | AH1 |
| 103 109 | FF1 | 103 139 | FF1 | 103 167 | FF1 | 103 195 | AH1 | 103 226* | AH1 |
| 103 110 | FF1 | 103 140 | FF1 | 103 168 | FF1 | 103 196 | AH1 | 103 227* | AH1 |
| 103 111 | FF1 | 103 141 | FF1 | 103 169 | FF1 | 103 197 | AH1 | 103 228* | AH1 |
| 103 112 | FF1 | 103 142 | FF1 | 103 170 | FF1 | 103 199 | AH1 | 103 229* | AH1 |
| 103 113 | FF1 | 103 143 | FF1 | 103 171 | FF1 | 103 200 | AH1 | 103 230* | AH1 |
| 103 114 | FF1 | 103 144 | FF1 | 103 172 | FF1 | 103 201 | AH1 | 103 231* | AH1 |
| 103 115 | FF1 | 103 145 | FF1 | 103 173* | AH1 | 103 202 | AH1 | 103 232* | AH1 |
| 103 116 | FF1 | 103 146 | FF1 | 103 174 | FF1 | 103 203 | AH1 | 103 233* | AH1 |
| 103 117 | FF1 | 103 147 | FF1 | 103 175 | AH1 | 103 204 | AH1 | 103 234* | AH1 |
| 103 118 | FF1 | 103 148 | FF1 | 103 176 | AH1 | 103 205 | AH1 | 103 235* | AH1 |
| 103 119 | FF1 | 103 149 | FF1 | 103 177 | AH1 | 103 206 | AH1 | 103 236* | AH1 |
| 103 120 | FF1 | 103 150 | FF1 | 103 178 | AH1 | 103 207 | AH1 | 103 237* | AH1 |
| 103 121 | FF1 | 103 151 | FF1 | 103 179 | AH1 | 103 208 | AH1 | 103 238* | AH1 |
| 103 122 | FF1 | 103 152 | FF1 | 103 180 | AH1 | 103 209 | AH1 | 103 239* | AH1 |
| 103 123 | FF1 | 103 153 | FF1 | 103 181 | AH1 | 103 210 | AH1 | 103 240* | AH1 |
| 103 124 | FF1 | 103 154 | FF1 | 103 182 | AH1 | 103 212 | AH1 | 103 241* | AH1 |
| 103 126 | FF1 | 103 155 | FF1 | 103 183 | AH1 | 103 213 | AH1 | 103 242* | AH1 |
| 103 127 | FF1 | 103 156 | FF1 | 103 184 | AH1 | 103 214 | AH1 | 103 243* | AH1 |
| 103 128 | FF1 | 103 157 | FF1 | 103 185 | AH1 | 103 215 | AH1 | 103 244* | AH1 |
| 103 129 | FF1 | 103 158 | FF1 | 103 186 | AH1 | 103 216* | AH1 | 103 245* | AH1 |
| 103 130 | FF1 | 103 159 | FF1 | 103 187 | AH1 | 103 217* | AH1 | | |

## CLASS 109    Bo-Bo

**Built:** 1961-76 by LEW.
**Electrical Equipment:** LEW.  **Continuous Rating:** 2740 kW.
**Weight:** 82.50 tonnes.  **Maximum Speed:** 120 km/h.
**Length:** 16.260 m.  **Train Supply:** Electric.
**Note:** Previously DR Class 211.

109 043 LH2    109 048  LH2    109 070  LH2    109 089  LH2

## CLASS 110    Bo-Bo

**Built:** 1956-69 by Henschel/Krauss-Maffei/Krupp.
**Electrical Equipment:** BBC/Siemens/AEG.  **Continuous Rating:** 3620 kW.

# GERMANY

**Weight:** 84.60 tonnes.  **Maximum Speed:** 150 km/h
**Length:** 16.490 m.  **Train Supply:** Electric.
**Note:** 110 485-504 were converted from 114 485-504 in 1991-95.

| | | | | | | | | | |
|---|---|---|---|---|---|---|---|---|---|
| 110 101 | FF1 | 110 167 | TS | 110 231 | TS | 110 305 | KK | 110 370 | AH1 |
| 110 102 | FF1 | 110 168 | FF1 | 110 232 | TS | 110 306 | KK | 110 371 | AH1 |
| 110 103 | FF1 | 110 169 | AH1 | 110 233 | TS | 110 307 | KK | 110 372 | AH1 |
| 110 105 | FF1 | 110 170 | TS | 110 234 | TS | 110 314 | KK | 110 373 | AH1 |
| 110 106 | FF1 | 110 171 | TS | 110 235 | TS | 110 315 | KK | 110 374 | AH1 |
| 110 107 | FF1 | 110 173 | TS | 110 236 | TS | 110 316 | KK | 110 375 | AH1 |
| 110 108 | FF1 | 110 174 | TS | 110 237 | TS | 110 317 | KK | 110 376 | AH1 |
| 110 109 | FF1 | 110 175 | TS | 110 238 | TS | 110 318 | KK | 110 377 | AH1 |
| 110 110 | FF1 | 110 176 | TS | 110 239 | TS | 110 319 | KK | 110 378 | AH1 |
| 110 111 | FF1 | 110 178 | TS | 110 240 | TS | 110 320 | KK | 110 379 | AH1 |
| 110 112 | FF1 | 110 179 | TS | 110 241 | TS | 110 321 | AH1 | 110 380 | AH1 |
| 110 113 | FF1 | 110 180 | TS | 110 242 | EDO | 110 322 | AH1 | 110 381 | AH1 |
| 110 114 | FF1 | 110 181 | TS | 110 243 | EDO | 110 323 | AH1 | 110 382 | AH1 |
| 110 115 | FF1 | 110 182 | TS | 110 244 | EDO | 110 324 | AH1 | 110 383 | AH1 |
| 110 116 | FF1 | 110 183 | TS | 110 245 | EDO | 110 325 | AH1 | 110 384 | AH1 |
| 110 117 | KK | 110 184 | TS | 110 247 | EDO | 110 326 | AH1 | 110 386 | FF1 |
| 110 118 | KK | 110 185 | TS | 110 248 | FF1 | 110 327 | AH1 | 110 387 | FF1 |
| 110 119 | KK | 110 186 | TS | 110 249 | EDO | 110 328 | AH1 | 110 388 | FF1 |
| 110 120 | KK | 110 187 | TS | 110 251 | EDO | 110 329 | AH1 | 110 389 | FF1 |
| 110 121 | KK | 110 188 | TS | 110 252 | EDO | 110 330 | AH1 | 110 391 | FF1 |
| 110 123 | KK | 110 189 | TS | 110 253 | KK | 110 331 | AH1 | 110 392 | FF1 |
| 110 124 | KK | 110 190 | TS | 110 254 | KK | 110 332 | FF1 | 110 393 | EDO |
| 110 125 | KK | 110 191 | TS | 110 256 | KK | 110 333 | FF1 | 110 394 | EDO |
| 110 126 | KK | 110 192 | TS | 110 257 | KK | 110 334 | FF1 | 110 395 | EDO |
| 110 127 | KK | 110 193 | TS | 110 258 | KK | 110 335 | FF1 | 110 396 | EDO |
| 110 128 | KK | 110 194 | TS | 110 259 | KK | 110 336 | FF1 | 110 397 | EDO |
| 110 129 | KK | 110 195 | TS | 110 261 | KK | 110 337 | FF1 | 110 398 | EDO |
| 110 130 | KK | 110 196 | TS | 110 263 | KK | 110 338 | FF1 | 110 399 | EDO |
| 110 131 | KK | 110 197 | TS | 110 271 | FF1 | 110 339 | FF1 | 110 400 | EDO |
| 110 132 | KK | 110 198 | TS | 110 272 | KK | 110 340 | FF1 | 110 401 | EDO |
| 110 133 | KK | 110 199 | KK | 110 273 | KK | 110 341 | FF1 | 110 402 | EDO |
| 110 134 | KK | 110 200 | KK | 110 274 | KK | 110 342 | FF1 | 110 403 | FF1 |
| 110 135 | KK | 110 201 | EDO | 110 275 | KK | 110 343 | FF1 | 110 404 | FF1 |
| 110 137 | KK | 110 202 | EDO | 110 276 | KK | 110 344 | KK | 110 405 | FF1 |
| 110 138 | KK | 110 203 | EDO | 110 277 | KK | 110 345 | FF1 | 110 406 | EDO |
| 110 140 | KK | 110 204 | EDO | 110 278 | KK | 110 346 | FF1 | 110 407 | EDO |
| 110 141 | KK | 110 205 | FF1 | 110 279 | KK | 110 347 | FF1 | 110 408 | EDO |
| 110 142 | KK | 110 206 | EDO | 110 280 | KK | 110 348 | FF1 | 110 409 | EDO |
| 110 143 | KK | 110 207 | FF1 | 110 281 | KK | 110 349 | FF1 | 110 410 | FF1 |
| 110 144 | KK | 110 208 | FF1 | 110 282 | KK | 110 350 | AH1 | 110 411 | EDO |
| 110 146 | KK | 110 209 | EDO | 110 284 | AH1 | 110 351 | FF1 | 110 412 | EDO |
| 110 147 | KK | 110 210 | FF1 | 110 286 | AH1 | 110 352 | KK | 110 413 | EDO |
| 110 148 | KK | 110 211 | FF1 | 110 288 | FF1 | 110 353 | KK | 110 414 | EDO |
| 110 149 | KK | 110 212 | FF1 | 110 289 | FF1 | 110 354 | KK | 110 415 | EDO |
| 110 150 | KK | 110 215 | FF1 | 110 290 | TS | 110 355 | AH1 | 110 416 | EDO |
| 110 151 | KK | 110 216 | TS | 110 291 | TS | 110 356 | AH1 | 110 417 | EDO |
| 110 152 | KK | 110 217 | TS | 110 292 | AH1 | 110 357 | AH1 | 110 418 | AH1 |
| 110 153 | KK | 110 218 | TS | 110 293 | TS | 110 358 | AH1 | 110 419 | EDO |
| 110 154 | FF1 | 110 219 | TS | 110 294 | TS | 110 359 | AH1 | 110 420 | EDO |
| 110 155 | KK | 110 221 | TS | 110 295 | TS | 110 360 | AH1 | 110 421 | EDO |
| 110 156 | KK | 110 222 | TS | 110 296 | AH1 | 110 361 | AH1 | 110 423 | EDO |
| 110 158 | KK | 110 223 | TS | 110 297 | FF1 | 110 362 | AH1 | 110 424 | EDO |
| 110 159 | KK | 110 224 | TS | 110 298 | AH1 | 110 363 | AH1 | 110 425 | EDO |
| 110 160 | KK | 110 225 | TS | 110 299 | AH1 | 110 364 | AH1 | 110 426 | EDO |
| 110 161 | KK | 110 226 | TS | 110 300 | AH1 | 110 365 | AH1 | 110 427 | AH1 |
| 110 162 | KK | 110 227 | TS | 110 301 | AH1 | 110 366 | AH1 | 110 428 | EDO |
| 110 164 | FF1 | 110 228 | TS | 110 302 | AH1 | 110 367 | AH1 | 110 429 | AH1 |
| 110 165 | FF1 | 110 229 | TS | 110 303 | AH1 | 110 368 | AH1 | 110 430 | EDO |
| 110 166 | TS | 110 230 | TS | 110 304 | KK | 110 369 | AH1 | 110 431 | EDO |

| | | | | | | | | | |
|---|---|---|---|---|---|---|---|---|---|
| 110 432 | EDO | 110 449 | EDO | 110 465 | EDO | 110 482 | EDO | 110 497 | TS |
| 110 434 | EDO | 110 450 | EDO | 110 466 | EDO | 110 483 | EDO | 110 498 | TS |
| 110 435 | EDO | 110 451 | EDO | 110 467 | EDO | 110 484 | EDO | 110 499 | FF1 |
| 110 436 | AH1 | 110 452 | EDO | 110 468 | EDO | 110 485 | AH1 | 110 500 | EDO |
| 110 437 | EDO | 110 453 | EDO | 110 469 | EDO | 110 486 | KK | 110 501 | EDO |
| 110 438 | EDO | 110 454 | EDO | 110 470 | EDO | 110 487 | KK | 110 502 | AH1 |
| 110 439 | EDO | 110 455 | EDO | 110 471 | EDO | 110 488 | KK | 110 503 | AH1 |
| 110 440 | EDO | 110 456 | EDO | 110 472 | EDO | 110 489 | AH1 | 110 504 | EDO |
| 110 441 | EDO | 110 457 | EDO | 110 473 | EDO | 110 490 | KK | 110 505 | EDO |
| 110 442 | AH1 | 110 458 | EDO | 110 474 | EDO | 110 491 | AH1 | 110 506 | EDO |
| 110 443 | EDO | 110 459 | EDO | 110 475 | EDO | 110 492 | KK | 110 507 | EDO |
| 110 444 | AH1 | 110 460 | EDO | 110 476 | EDO | 110 493 | AH1 | 110 508 | EDO |
| 110 445 | EDO | 110 461 | EDO | 110 478 | EDO | 110 494 | AH1 | 110 509 | EDO |
| 110 446 | EDO | 110 462 | EDO | 110 479 | EDO | 110 495 | AH1 | 110 510 | EDO |
| 110 447 | EDO | 110 463 | EDO | 110 480 | EDO | 110 496 | AH1 | 110 511 | EDO |
| 110 448 | EDO | 110 464 | EDO | 110 481 | EDO | | | | |

# CLASS 111        Bo-Bo

**Built:** 1975-84 by Henschel/Krauss-Maffei/Krupp.
**Electrical Equipment:** BBC/Siemens/AEG.
**Weight:** 83.00 tonnes.
**Length:** 16.750 m.
**Continuous Rating:** 3850 kW.
**Maximum Speed:** 160 km/h.
**Train Supply:** Electric.

| | | | | | | | | | |
|---|---|---|---|---|---|---|---|---|---|
| 111 001 | MH1 | 111 040 | MH1 | 111 079 | MH1 | 111 119 | EDO | 111 158 | KD |
| 111 002 | MH1 | 111 041 | MH1 | 111 080 | MH1 | 111 120 | EDO | 111 159 | KD |
| 111 003 | MH1 | 111 042 | MH1 | 111 081 | MH1 | 111 121 | KD | 111 160 | KD |
| 111 004 | MH1 | 111 043 | MH1 | 111 082 | FF1 | 111 122 | EDO | 111 161 | KD |
| 111 005 | MH1 | 111 044 | MH1 | 111 083 | FF1 | 111 123 | EDO | 111 162 | KD |
| 111 006 | MH1 | 111 045 | MH1 | 111 084 | FF1 | 111 124 | EDO | 111 163 | KD |
| 111 007 | MH1 | 111 046 | MH1 | 111 085 | FF1 | 111 125 | EDO | 111 164 | KD |
| 111 008 | MH1 | 111 047 | MH1 | 111 086 | FF1 | 111 126 | EDO | 111 165 | KD |
| 111 009 | MH1 | 111 048 | MH1 | 111 087 | FF1 | 111 127 | EDO | 111 166 | KD |
| 111 010 | MH1 | 111 049 | MH1 | 111 088 | FF1 | 111 128 | EDO | 111 167 | KD |
| 111 011 | MH1 | 111 050 | MH1 | 111 089 | FF1 | 111 129 | EDO | 111 168 | KD |
| 111 012 | MH1 | 111 051 | MH1 | 111 090 | FF1 | 111 130 | KD | 111 169 | KD |
| 111 013 | MH1 | 111 052 | MH1 | 111 091 | FF1 | 111 131 | EDO | 111 170 | KD |
| 111 014 | MH1 | 111 053 | MH1 | 111 092 | FF1 | 111 132 | EDO | 111 171 | KD |
| 111 015 | MH1 | 111 054 | MH1 | 111 093 | FF1 | 111 133 | EDO | 111 172 | KD |
| 111 016 | MH1 | 111 055 | MH1 | 111 094 | FF1 | 111 134 | EDO | 111 173 | KD |
| 111 017 | MH1 | 111 056 | MH1 | 111 095 | FF1 | 111 135 | EDO | 111 174 | KD |
| 111 018 | MH1 | 111 057 | FF1 | 111 096 | FF1 | 111 136 | EDO | 111 175 | KD |
| 111 019 | MH1 | 111 058 | FF1 | 111 097 | FF1 | 111 137 | EDO | 111 176 | KD |
| 111 020 | MH1 | 111 059 | MH1 | 111 098 | FF1 | 111 138 | KD | 111 177 | MH1 |
| 111 021 | MH1 | 111 060 | FF1 | 111 099 | FF1 | 111 139 | EDO | 111 178 | MH1 |
| 111 022 | MH1 | 111 061 | FF1 | 111 100 | FF1 | 111 140 | EDO | 111 179 | MH1 |
| 111 023 | MH1 | 111 062 | MH1 | 111 101 | FF1 | 111 141 | KD | 111 180 | MH1 |
| 111 024 | MH1 | 111 063 | MH1 | 111 102 | FF1 | 111 142 | KD | 111 181 | MH1 |
| 111 025 | MH1 | 111 064 | MH1 | 111 103 | FF1 | 111 143 | KD | 111 182 | MH1 |
| 111 026 | MH1 | 111 065 | MH1 | 111 104 | FF1 | 111 144 | KD | 111 183 | MH1 |
| 111 027 | MH1 | 111 066 | MH1 | 111 105 | FF1 | 111 145 | EDO | 111 184 | MH1 |
| 111 028 | MH1 | 111 067 | MH1 | 111 106 | FF1 | 111 146 | KD | 111 185 | MH1 |
| 111 029 | MH1 | 111 068 | MH1 | 111 107 | FF1 | 111 147 | KD | 111 186 | MH1 |
| 111 030 | MH1 | 111 069 | MH1 | 111 108 | FF1 | 111 148 | KD | 111 187 | MH1 |
| 111 031 | MH1 | 111 070 | MH1 | 111 110 | FF1 | 111 149 | KD | 111 188 | MH1 |
| 111 032 | MH1 | 111 071 | MH1 | 111 111 | KD | 111 150 | KD | 111 189 | MH1 |
| 111 033 | MH1 | 111 072 | MH1 | 111 112 | KD | 111 151 | KD | 111 190 | MH1 |
| 111 034 | MH1 | 111 073 | MH1 | 111 113 | EDO | 111 152 | KD | 111 191 | MH1 |
| 111 035 | MH1 | 111 074 | MH1 | 111 114 | KD | 111 153 | KD | 111 192 | MH1 |
| 111 036 | MH1 | 111 075 | MH1 | 111 115 | EDO | 111 154 | KD | 111 193 | MH1 |
| 111 037 | MH1 | 111 076 | MH1 | 111 116 | EDO | 111 155 | KD | 111 194 | MH1 |
| 111 038 | MH1 | 111 077 | MH1 | 111 117 | EDO | 111 156 | KD | 111 195 | MH1 |
| 111 039 | MH1 | 111 078 | MH1 | 111 118 | EDO | 111 157 | KD | 111 196 | MH1 |

# GERMANY

◄ 110 156 at Remagen with the 1327 Koblenz - Köln Deutz on 14.01.92 (M. Dunn)

▼ 111 177 at Garmisch-Partenkirchen with the 1300 Mittenwald - München on 18.04.94 (D. Guy)

◄ 099 723 at Seifersdorf on 10.02.93 (J. Hayes)

▼ 103 163 at Hamburg Dammtor with the 0659 Aachen - Kiel on 27.02.93 (M. Dunn)

| | | | | | | | | |
|---|---|---|---|---|---|---|---|---|
| 111 197 | MH1 | 111 204 | MH1 | 111 210 | FF1 | 111 216 | MH1 | 111 222 | MH1 |
| 111 198 | MH1 | 111 205 | MH1 | 111 211 | MH1 | 111 217 | MH1 | 111 223 | MH1 |
| 111 199 | MH1 | 111 206 | MH1 | 111 212 | FF1 | 111 218 | FF1 | 111 224 | MH1 |
| 111 200 | MH1 | 111 207 | MH1 | 111 213 | MH1 | 111 219 | MH1 | 111 225 | MH1 |
| 111 201 | MH1 | 111 208 | MH1 | 111 214 | MH1 | 111 220 | MH1 | 111 226 | MH1 |
| 111 202 | MH1 | 111 209 | MH1 | 111 215 | MH1 | 111 221 | MH1 | 111 227 | MH1 |
| 111 203 | MH1 | | | | | | | | |

## CLASS 112                                                       Bo-Bo

**Built:** 1990-94.
**Electrical Equipment:** LEW.                  **Continuous Rating:** 3500 kW.
**Weight:** 82.50 tonnes.                    **Maximum Speed:** 140 km/h.
**Length:** 16.640 m.                       **Train Supply:** Electric.
**Note:** Previously DR Class 212.

| | | | | | | | | | |
|---|---|---|---|---|---|---|---|---|---|
| 112 002 | BHF | 112 029 | BHF | 112 115 | BHF | 112 141 | BHF | 112 166 | BHF |
| 112 003 | BHF | 112 030 | BHF | 112 116 | BHF | 112 142 | BHF | 112 167 | BHF |
| 112 004 | BHF | 112 031 | BHF | 112 117 | BHF | 112 143 | BHF | 112 168 | BHF |
| 112 005 | BHF | 112 032 | BHF | 112 118 | BHF | 112 144 | BHF | 112 169 | BHF |
| 112 006 | BHF | 112 033 | BHF | 112 119 | BHF | 112 145 | BHF | 112 170 | BHF |
| 112 007 | BHF | 112 034 | BHF | 112 120 | BHF | 112 146 | BHF | 112 171 | BHF |
| 112 008 | BHF | 112 035 | BHF | 112 121 | BHF | 112 147 | BHF | 112 172 | BHF |
| 112 009 | BHF | 112 036 | BHF | 112 122 | BHF | 112 148 | BHF | 112 173 | BHF |
| 112 010 | BHF | 112 037 | BHF | 112 123 | BHF | 112 149 | BHF | 112 174 | BHF |
| 112 011 | BHF | 112 038 | BHF | 112 124 | BHF | 112 150 | BHF | 112 175 | BHF |
| 112 012 | BHF | 112 039 | BHF | 112 125 | BHF | 112 151 | BHF | 112 176 | BHF |
| 112 013 | BHF | 112 040 | BHF | 112 126 | BHF | 112 152 | BHF | 112 177 | BHF |
| 112 014 | BHF | 112 101 | BHF | 112 127 | BHF | 112 153 | BHF | 112 178 | BHF |
| 112 015 | BHF | 112 102 | BHF | 112 128 | BHF | 112 154 | BHF | 112 179 | BHF |
| 112 016 | BHF | 112 103 | BHF | 112 129 | BHF | 112 155 | BHF | 112 180 | BHF |
| 112 017 | BHF | 112 104 | BHF | 112 130 | BHF | 112 156 | BHF | 112 181 | BHF |
| 112 018 | BHF | 112 105 | BHF | 112 131 | BHF | 112 157 | BHF | 112 182 | BHF |
| 112 019 | BHF | 112 106 | BHF | 112 132 | BHF | 112 158 | BHF | 112 183 | BHF |
| 112 020 | BHF | 112 107 | BHF | 112 133 | BHF | 112 159 | BHF | 112 184 | BHF |
| 112 021 | BHF | 112 108 | BHF | 112 134 | BHF | 112 160 | BHF | 112 185 | BHF |
| 112 022 | BHF | 112 109 | BHF | 112 135 | BHF | 112 161 | BHF | 112 186 | BHF |
| 112 023 | BHF | 112 110 | BHF | 112 136 | BHF | 112 162 | BHF | 112 187 | BHF |
| 112 024 | BHF | 112 111 | BHF | 112 137 | BHF | 112 163 | BHF | 112 188 | BHF |
| 112 026 | BHF | 112 112 | BHF | 112 138 | BHF | 112 164 | BHF | 112 189 | BHF |
| 112 027 | BHF | 112 113 | BHF | 112 139 | BHF | 112 165 | BHF | 112 190 | BHF |
| 112 028 | BHF | 112 114 | BHF | 112 140 | BHF | | | | |

## CLASS 113                                                       Bo-Bo

**Built:** 1962-64 by Krauss-Maffei as class 110. Reclassified 1968.
**Electrical Equipment:** Siemens.             **Continuous Rating:** 3620 kW .
**Weight:** 86.00 tonnes.                    **Maximum Speed:** 160 km/h.
**Length:** 16.490 m.                       **Train Supply:** Electric.

| | | | | | | | | | |
|---|---|---|---|---|---|---|---|---|---|
| 113 265 | MH1 | 113 268 | MH1 | 113 270 | MH1 | 113 309 | MH1 | 113 311 | MH1 |
| 113 266 | MH1 | 113 269 | MH1 | 113 308 | MH1 | 113 310 | MH1 | 113 312 | MH1 |
| 113 267 | MH1 | | | | | | | | |

## CLASS 120                                                       Bo-Bo

**Built:** 1979, 1987-89 by Henschel/Krauss-Maffei/Krupp.
**Electrical Equipment:** BBC.                 **Continuous Rating:** 5600 kW.
**Weight:** 83.20 tonnes.                    **Maximum Speed:** 200 km/h.
**Length:** 19.200 m.                       **Train Supply:** Electric.

| | | | | | | | | | |
|---|---|---|---|---|---|---|---|---|---|
| 120 002 | NN2 | 120 005 | NN2 | 120 103 | NN2 | 120 106 | NN2 | 120 109 | NN2 |
| 120 003 | NN2 | 120 101 | NN2 | 120 104 | NN2 | 120 107 | NN2 | 120 110 | NN2 |
| 120 004 | NN2 | 120 102 | NN2 | 120 105 | NN2 | 120 108 | NN2 | 120 111 | NN2 |

# GERMANY

| | | | | | | | | | |
|---|---|---|---|---|---|---|---|---|---|
| 120 112 | NN2 | 120 122 | NN2 | 120 132 | NN2 | 120 142 | NN2 | 120 152 | NN2 |
| 120 113 | NN2 | 120 123 | NN2 | 120 133 | NN2 | 120 143 | NN2 | 120 153 | NN2 |
| 120 114 | NN2 | 120 124 | NN2 | 120 134 | NN2 | 120 144 | NN2 | 120 154 | NN2 |
| 120 115 | NN2 | 120 125 | NN2 | 120 135 | NN2 | 120 145 | NN2 | 120 155 | NN2 |
| 120 116 | NN2 | 120 126 | NN2 | 120 136 | NN2 | 120 146 | NN2 | 120 156 | NN2 |
| 120 117 | NN2 | 120 127 | NN2 | 120 137 | NN2 | 120 147 | NN2 | 120 157 | NN2 |
| 120 118 | NN2 | 120 128 | NN2 | 120 138 | NN2 | 120 148 | NN2 | 120 158 | NN2 |
| 120 119 | NN2 | 120 129 | NN2 | 120 139 | NN2 | 120 149 | NN2 | 120 159 | NN2 |
| 120 120 | NN2 | 120 130 | NN2 | 120 140 | NN2 | 120 150 | NN2 | 120 160 | NN2 |
| 120 121 | NN2 | 120 131 | NN2 | 120 141 | NN2 | 120 151 | NN2 | | |

**Name:** 120 002  *Fürth*

## CLASS 127                                                                Bo-Bo

**Built:** 1992 by Krauss-Maffei/Siemens.
**Electrical Equipment:** Siemens.                 **Continuous Rating:** 6400 kW.
**Weight:** 86.00 tonnes.                               **Maximum Speed:** 230 km/h.
**Length:** 19.580 m.                                      **Train Supply:** Electric.
**Note:** Test locomotive owned by Siemens/Krauss Maffei on loan to DBAG.

127 001  MH1

## CLASS 128                                                                Bo-Bo

**Built:** 1994 by AEG.
**Electrical Equipment:** AEG.                        **Continuous Rating:** 6400 kW.
**Weight:** 84.00 tonnes.                               **Maximum Speed:** 220 km/h.
**Length:** 19.500 m.                                      **Train Supply:** Electric.
**Note:** Test locomotive owned by AEG on loan to DBAG.

128 001  MH1

## CLASS 139                                                                Bo-Bo

**Built:** 1959-65 by Henschel/Krauss-Maffei/Krupp.
**Electrical Equipment:** BBC/Siemens/AEG.     **Continuous Rating:** 3620 kW.
**Weight:** 84.60 tonnes.                               **Maximum Speed:** 110 km/h.
**Length:** 16.490 m.                                      **Train Supply:** Electric.
**Note:** Some locomotives have recently been renumbered from Class 110, retaining their original serial numbers.

| | | | | | | | | | |
|---|---|---|---|---|---|---|---|---|---|
| 139 122 | MH1 | 139 163 | MH1 | 139 250 | MH1 | 139 310 | MH1 | 139 554 | MH1 |
| 139 131 | MH1 | 139 164 | MH1 | 139 255 | MH1 | 139 311 | MH1 | 139 555 | MH1 |
| 139 132 | MH1 | 139 165 | MH1 | 139 260 | MH1 | 139 312 | RM | 139 556 | MH1 |
| 139 133 | MH1 | 139 166 | MH1 | 139 262 | MH1 | 139 313 | RM | 139 557 | MH1 |
| 139 135 | MH1 | 139 172 | MH1 | 139 264 | MH1 | 139 314 | RM | 139 558 | MH1 |
| 139 136 | RM | 139 177 | MH1 | 139 283 | MH1 | 139 315 | RM | 139 559 | MH1 |
| 139 137 | MH1 | 139 213 | MH1 | 139 285 | MH1 | 139 316 | RM | 139 560 | MH1 |
| 139 139 | MH1 | 139 214 | MH1 | 139 287 | MH1 | 139 552 | MH1 | 139 561 | MH1 |
| 139 145 | MH1 | 139 220 | MH1 | 139 309 | MH1 | 139 553 | MH1 | 139 562 | MH1 |
| 139 157 | MH1 | 139 246 | MH1 | | | | | | |

## CLASS 140                                                                Bo-Bo

**Built:** 1957-73 by Henschel/Krauss-Maffei/Krupp.
**Electrical Equipment:** BBC/Siemens/AEG.     **Continuous Rating:** 3620 kW.
**Weight:** 83.00 tonnes.                               **Maximum Speed:** 110 km/h.
**Length:** 16.490 m.                                      **Train Supply:** Electric.

| | | | | | | | | | |
|---|---|---|---|---|---|---|---|---|---|
| 140 001 | MH1 | 140 008 | MH1 | 140 014 | MH1 | 140 020 | MH1 | 140 026 | MH1 |
| 140 002 | MH1 | 140 009 | MH1 | 140 015 | MH1 | 140 021 | MH1 | 140 027 | MH1 |
| 140 003 | MH1 | 140 010 | MH1 | 140 016 | MH1 | 140 022 | MH1 | 140 028 | MH1 |
| 140 004 | MH1 | 140 011 | MH1 | 140 017 | MH1 | 140 023 | MH1 | 140 029 | MH1 |
| 140 005 | MH1 | 140 012 | MH1 | 140 018 | MH1 | 140 024 | MH1 | 140 030 | MH1 |
| 140 006 | MH1 | 140 013 | MH1 | 140 019 | MH1 | 140 025 | MH1 | 140 031 | MH1 |

| | | | | | | | | | |
|---|---|---|---|---|---|---|---|---|---|
| 140 032 | MH1 | 140 110 | AH1 | 140 199 | KK | 140 269 | KK | 140 346 | KK |
| 140 033 | MH1 | 140 111 | AH1 | 140 200 | KK | 140 270 | EDO | 140 347 | KK |
| 140 034 | MH1 | 140 112 | AH1 | 140 201 | KK | 140 271 | RM | 140 348 | KK |
| 140 035 | MH1 | 140 113 | AH1 | 140 202 | KK | 140 272 | EDO | 140 349 | KK |
| 140 036 | MH1 | 140 114 | AH1 | 140 203 | KK | 140 273 | EDO | 140 350 | RM |
| 140 037 | MH1 | 140 115 | RM | 140 204 | KK | 140 274 | EDO | 140 351 | KK |
| 140 038 | MH1 | 140 116 | HS | 140 205 | KK | 140 275 | EDO | 140 352 | KK |
| 140 039 | MH1 | 140 117 | HS | 140 206 | KK | 140 276 | EDO | 140 353 | KK |
| 140 040 | MH1 | 140 118 | KK | 140 207 | KK | 140 278 | EDO | 140 354 | KK |
| 140 041 | MH1 | 140 119 | KK | 140 208 | KK | 140 279 | EDO | 140 355 | RM |
| 140 042 | MH1 | 140 120 | KK | 140 209 | KK | 140 280 | EDO | 140 356 | RM |
| 140 043 | MH1 | 140 122 | KK | 140 210 | KK | 140 281 | EDO | 140 357 | KK |
| 140 044 | MH1 | 140 123 | KK | 140 211 | RM | 140 282 | EDO | 140 358 | KK |
| 140 045 | MH1 | 140 124 | EDO | 140 212 | RM | 140 283 | EDO | 140 359 | KK |
| 140 046 | MH1 | 140 125 | KK | 140 213 | RM | 140 284 | EDO | 140 360 | RM |
| 140 047 | MH1 | 140 126 | KK | 140 214 | RM | 140 285 | EDO | 140 361 | RM |
| 140 048 | EDO | 140 127 | KK | 140 215 | RM | 140 286 | EDO | 140 362 | HS |
| 140 050 | SSH | 140 128 | KK | 140 216 | RM | 140 287 | EDO | 140 363 | RM |
| 140 051 | KK | 140 129 | KK | 140 217 | RM | 140 288 | EDO | 140 364 | RM |
| 140 052 | EDO | 140 130 | KK | 140 218 | RM | 140 289 | EDO | 140 365 | RM |
| 140 053 | MH1 | 140 138 | HS | 140 219 | RM | 140 290 | HS | 140 366 | RM |
| 140 054 | MH1 | 140 139 | SSH | 140 220 | RM | 140 291 | KK | 140 367 | HS |
| 140 055 | SSH | 140 143 | KK | 140 221 | RM | 140 292 | KK | 140 368 | HS |
| 140 056 | SSH | 140 145 | SSH | 140 222 | RM | 140 293 | KK | 140 369 | HS |
| 140 057 | MH1 | 140 146 | SSH | 140 223 | RM | 140 294 | EDO | 140 370 | HS |
| 140 058 | SSH | 140 147 | SSH | 140 224 | RM | 140 295 | FF1 | 140 371 | HS |
| 140 059 | MH1 | 140 148 | SSH | 140 225 | RM | 140 296 | EDO | 140 372 | HS |
| 140 060 | EDO | 140 149 | SSH | 140 227 | KK | 140 297 | FF1 | 140 373 | HS |
| 140 061 | SSH | 140 150 | RM | 140 228 | HS | 140 298 | SSH | 140 374 | HS |
| 140 062 | MH1 | 140 151 | RM | 140 229 | EDO | 140 299 | FF1 | 140 375 | HS |
| 140 063 | EDO | 140 153 | RM | 140 230 | KK | 140 300 | FF1 | 140 376 | HS |
| 140 064 | EDO | 140 154 | RM | 140 231 | KK | 140 303 | FF1 | 140 377 | HS |
| 140 065 | EDO | 140 156 | RM | 140 232 | FF1 | 140 304 | FF1 | 140 378 | HS |
| 140 067 | SSH | 140 157 | RM | 140 233 | FF1 | 140 305 | FF1 | 140 379 | HS |
| 140 069 | EDO | 140 159 | SSH | 140 234 | FF1 | 140 306 | HS | 140 380 | HS |
| 140 070 | FF1 | 140 160 | SSH | 140 235 | FF1 | 140 307 | HS | 140 381 | HS |
| 140 071 | FF1 | 140 161 | SSH | 140 236 | FF1 | 140 308 | FF1 | 140 382 | HS |
| 140 072 | EDO | 140 162 | RM | 140 238 | FF1 | 140 317 | KK | 140 383 | HS |
| 140 073 | KK | 140 168 | EDO | 140 239 | FF1 | 140 318 | KK | 140 384 | HS |
| 140 074 | EDO | 140 169 | RM | 140 240 | FF1 | 140 320 | KK | 140 385 | HS |
| 140 075 | EDO | 140 170 | RM | 140 241 | FF1 | 140 321 | KK | 140 386 | HS |
| 140 076 | EDO | 140 171 | RM | 140 243 | FF1 | 140 322 | KK | 140 387 | HS |
| 140 077 | MH1 | 140 172 | RM | 140 244 | FF1 | 140 323 | KK | 140 388 | HS |
| 140 078 | SSH | 140 173 | RM | 140 246 | HS | 140 324 | KK | 140 389 | HS |
| 140 079 | SSH | 140 176 | SSH | 140 247 | HS | 140 325 | KK | 140 390 | HS |
| 140 080 | MH1 | 140 178 | SSH | 140 248 | KK | 140 326 | KK | 140 391 | HS |
| 140 081 | MH1 | 140 179 | SSH | 140 249 | KK | 140 327 | KK | 140 392 | HS |
| 140 082 | SSH | 140 182 | RM | 140 251 | RM | 140 329 | KK | 140 393 | KK |
| 140 083 | MH1 | 140 183 | RM | 140 252 | RM | 140 330 | KK | 140 394 | KK |
| 140 084 | KK | 140 184 | RM | 140 253 | FF1 | 140 331 | KK | 140 395 | RM |
| 140 085 | KK | 140 185 | RM | 140 254 | FF1 | 140 332 | KK | 140 396 | RM |
| 140 091 | KK | 140 186 | RM | 140 255 | EDO | 140 333 | KK | 140 397 | RM |
| 140 095 | KK | 140 187 | KK | 140 258 | KK | 140 334 | KK | 140 398 | RM |
| 140 097 | SSH | 140 188 | KK | 140 259 | KK | 140 335 | KK | 140 399 | HS |
| 140 098 | SSH | 140 189 | KK | 140 260 | RM | 140 336 | KK | 140 400 | HS |
| 140 099 | SSH | 140 190 | KK | 140 261 | RM | 140 337 | KK | 140 401 | KK |
| 140 100 | EDO | 140 191 | SSH | 140 262 | KK | 140 338 | KK | 140 402 | KK |
| 140 101 | EDO | 140 192 | EDO | 140 263 | KK | 140 340 | KK | 140 403 | KK |
| 140 102 | SSH | 140 193 | EDO | 140 264 | KK | 140 341 | KK | 140 404 | KK |
| 140 103 | EDO | 140 194 | KK | 140 265 | EDO | 140 342 | KK | 140 405 | RM |
| 140 107 | AH1 | 140 195 | EDO | 140 266 | EDO | 140 343 | KK | 140 406 | RM |
| 140 108 | AH1 | 140 196 | EDO | 140 267 | EDO | 140 344 | KK | 140 407 | RM |
| 140 109 | AH1 | 140 197 | RM | 140 268 | KK | 140 345 | KK | 140 408 | RM |

# GERMANY

| | | | | | | | | | |
|---|---|---|---|---|---|---|---|---|---|
| 140 409 | HS | 140 473 | SSH | 140 538 | HS | 140 613 | HS | 140 677 | RM |
| 140 410 | HS | 140 474 | SSH | 140 539 | HS | 140 614 | HS | 140 678 | RM |
| 140 411 | HS | 140 475 | SSH | 140 540 | HS | 140 615 | HS | 140 679 | RM |
| 140 412 | HS | 140 476 | SSH | 140 541 | HS | 140 616 | HS | 140 680 | RM |
| 140 413 | HS | 140 477 | SSH | 140 542 | HS | 140 617 | HS | 140 681 | RM |
| 140 414 | HS | 140 478 | SSH | 140 543 | HS | 140 618 | HS | 140 682 | RM |
| 140 415 | EDO | 140 479 | SSH | 140 544 | HS | 140 619 | HS | 140 683 | AH1 |
| 140 416 | EDO | 140 480 | SSH | 140 545 | HS | 140 620 | HS | 140 684 | KK |
| 140 417 | EDO | 140 481 | SSH | 140 546 | FF1 | 140 621 | FF1 | 140 685 | HS |
| 140 419 | HS | 140 482 | RM | 140 547 | HS | 140 622 | FF1 | 140 686 | HS |
| 140 420 | HS | 140 483 | RM | 140 548 | HS | 140 623 | FF1 | 140 687 | HS |
| 140 421 | HS | 140 484 | RM | 140 549 | HS | 140 624 | HS | 140 688 | HS |
| 140 422 | HS | 140 486 | RM | 140 550 | HS | 140 625 | FF1 | 140 689 | HO |
| 140 423 | HS | 140 487 | RM | 140 551 | HS | 140 626 | FF1 | 140 690 | KK |
| 140 424 | HS | 140 488 | RM | 140 564 | RM | 140 627 | FF1 | 140 691 | HS |
| 140 425 | HS | 140 489 | RM | 140 565 | RM | 140 628 | HS | 140 692 | HS |
| 140 426 | HS | 140 490 | RM | 140 566 | RM | 140 629 | FF1 | 140 693 | HO |
| 140 427 | HS | 140 491 | RM | 140 567 | RM | 140 630 | HS | 140 694 | HO |
| 140 428 | HS | 140 492 | RM | 140 568 | RM | 140 631 | KK | 140 695 | HO |
| 140 429 | HS | 140 493 | RM | 140 569 | RM | 140 632 | KK | 140 696 | HO |
| 140 430 | FF1 | 140 494 | RM | 140 570 | RM | 140 634 | SSH | 140 697 | HO |
| 140 431 | FF1 | 140 495 | RM | 140 571 | RM | 140 635 | HS | 140 698 | HO |
| 140 432 | FF1 | 140 496 | RM | 140 572 | KK | 140 636 | HS | 140 700 | HO |
| 140 433 | HS | 140 497 | RM | 140 573 | KK | 140 637 | HS | 140 701 | HO |
| 140 434 | HS | 140 498 | RM | 140 574 | KK | 140 638 | HS | 140 702 | KK |
| 140 435 | HS | 140 499 | RM | 140 575 | KK | 140 639 | HS | 140 703 | AH1 |
| 140 436 | HS | 140 500 | RM | 140 576 | KK | 140 640 | HS | 140 704 | AH1 |
| 140 437 | HS | 140 501 | RM | 140 577 | EDO | 140 641 | HS | 140 705 | HO |
| 140 438 | HS | 140 502 | RM | 140 578 | EDO | 140 642 | HS | 140 706 | EDO |
| 140 439 | HS | 140 503 | RM | 140 579 | KK | 140 643 | SSH | 140 707 | HO |
| 140 440 | HS | 140 504 | RM | 140 580 | HS | 140 644 | KK | 140 708 | HO |
| 140 441 | HS | 140 505 | RM | 140 581 | HS | 140 645 | KK | 140 709 | KK |
| 140 442 | HS | 140 506 | RM | 140 582 | KK | 140 646 | KK | 140 710 | KK |
| 140 443 | HS | 140 507 | RM | 140 583 | EDO | 140 647 | HS | 140 711 | RM |
| 140 444 | HS | 140 508 | RM | 140 584 | KK | 140 648 | EDO | 140 712 | RM |
| 140 445 | HS | 140 509 | RM | 140 585 | FF1 | 140 649 | KK | 140 713 | HO |
| 140 446 | HS | 140 510 | RM | 140 586 | FF1 | 140 650 | KK | 140 714 | HO |
| 140 447 | RM | 140 511 | RM | 140 587 | FF1 | 140 651 | KK | 140 715 | HO |
| 140 448 | HS | 140 512 | RM | 140 588 | KK | 140 652 | KK | 140 716 | HO |
| 140 449 | HS | 140 513 | RM | 140 589 | KK | 140 653 | EDO | 140 717 | KK |
| 140 450 | HS | 140 514 | RM | 140 590 | KK | 140 654 | HS | 140 718 | KK |
| 140 451 | HS | 140 515 | RM | 140 591 | KK | 140 655 | EDO | 140 719 | AH1 |
| 140 452 | HS | 140 516 | RM | 140 592 | HS | 140 656 | EDO | 140 720 | AH1 |
| 140 453 | HS | 140 517 | RM | 140 593 | HS | 140 657 | EDO | 140 721 | HO |
| 140 454 | HS | 140 518 | RM | 140 594 | HS | 140 658 | EDO | 140 722 | HO |
| 140 455 | HS | 140 519 | RM | 140 595 | KK | 140 659 | EDO | 140 723 | SSH |
| 140 456 | HS | 140 520 | HS | 140 596 | KK | 140 660 | HS | 140 724 | SSH |
| 140 457 | HS | 140 521 | HS | 140 597 | KK | 140 661 | HS | 140 725 | HO |
| 140 458 | HS | 140 522 | HS | 140 598 | KK | 140 662 | HS | 140 726 | EDO |
| 140 459 | HS | 140 523 | HS | 140 599 | KK | 140 663 | HS | 140 727 | KK |
| 140 460 | RM | 140 524 | HS | 140 600 | KK | 140 664 | HS | 140 728 | KK |
| 140 461 | RM | 140 526 | HS | 140 601 | SSH | 140 665 | FF1 | 140 729 | AH1 |
| 140 462 | HS | 140 527 | HS | 140 602 | FF1 | 140 666 | SSH | 140 730 | AH1 |
| 140 463 | HS | 140 528 | HS | 140 603 | FF1 | 140 667 | FF1 | 140 731 | AH1 |
| 140 464 | HS | 140 529 | RM | 140 604 | FF1 | 140 668 | AH1 | 140 732 | AH1 |
| 140 465 | HS | 140 530 | RM | 140 605 | FF1 | 140 669 | AH1 | 140 733 | AH1 |
| 140 466 | HS | 140 531 | AH1 | 140 606 | FF1 | 140 670 | KK | 140 735 | AH1 |
| 140 467 | HS | 140 532 | HS | 140 607 | FF1 | 140 671 | KK | 140 736 | AH1 |
| 140 468 | HS | 140 533 | HS | 140 608 | AH1 | 140 672 | KK | 140 737 | HS |
| 140 469 | HS | 140 534 | HS | 140 609 | HS | 140 673 | KK | 140 738 | HS |
| 140 470 | HS | 140 535 | HS | 140 610 | HS | 140 674 | KK | 140 739 | EDO |
| 140 471 | SSH | 140 536 | HS | 140 611 | HS | 140 675 | KK | 140 740 | KK |
| 140 472 | SSH | 140 537 | HS | 140 612 | HS | 140 676 | KK | 140 741 | KK |

| | | | | | | | | | |
|---|---|---|---|---|---|---|---|---|---|
| 140 742 | KK | 140 770 | FF1 | 140 798 | FF1 | 140 826 | EDO | 140 853 | EDO |
| 140 743 | KK | 140 771 | HS | 140 799 | FF1 | 140 827 | EDO | 140 854 | EDO |
| 140 744 | EDO | 140 772 | RM | 140 800 | AH1 | 140 828 | EDO | 140 855 | EDO |
| 140 745 | EDO | 140 773 | HS | 140 801 | AH1 | 140 829 | EDO | 140 856 | EDO |
| 140 746 | EDO | 140 774 | HS | 140 802 | AH1 | 140 830 | EDO | 140 857 | EDO |
| 140 747 | EDO | 140 775 | HS | 140 803 | AH1 | 140 831 | EDO | 140 858 | EDO |
| 140 748 | AH1 | 140 776 | EDO | 140 804 | AH1 | 140 832 | EDO | 140 859 | EDO |
| 140 749 | FF1 | 140 777 | EDO | 140 805 | AH1 | 140 833 | KK | 140 860 | EDO |
| 140 750 | AH1 | 140 778 | HS | 140 806 | AH1 | 140 834 | KK | 140 861 | EDO |
| 140 751 | AH1 | 140 779 | HS | 140 807 | AH1 | 140 835 | KK | 140 862 | EDO |
| 140 752 | EDO | 140 780 | HS | 140 808 | AH1 | 140 836 | KK | 140 863 | EDO |
| 140 753 | KK | 140 781 | FF1 | 140 809 | AH1 | 140 837 | KK | 140 864 | EDO |
| 140 755 | KK | 140 782 | HS | 140 810 | AH1 | 140 838 | KK | 140 865 | SSH |
| 140 756 | KK | 140 783 | HS | 140 811 | AH1 | 140 839 | KK | 140 866 | SSH |
| 140 757 | FF1 | 140 784 | HS | 140 812 | FF1 | 140 840 | KK | 140 867 | EDO |
| 140 758 | HS | 140 785 | HS | 140 813 | FF1 | 140 841 | EDO | 140 868 | SSH |
| 140 759 | FF1 | 140 786 | FF1 | 140 814 | FF1 | 140 842 | EDO | 140 869 | SSH |
| 140 760 | FF1 | 140 787 | FF1 | 140 815 | SSH | 140 843 | EDO | 140 870 | SSH |
| 140 761 | FF1 | 140 788 | FF1 | 140 816 | EDO | 140 844 | EDO | 140 871 | SSH |
| 140 762 | FF1 | 140 789 | FF1 | 140 817 | EDO | 140 845 | EDO | 140 872 | SSH |
| 140 763 | FF1 | 140 790 | FF1 | 140 818 | EDO | 140 846 | EDO | 140 873 | SSH |
| 140 764 | RM | 140 791 | FF1 | 140 819 | EDO | 140 847 | EDO | 140 874 | SSH |
| 140 765 | FF1 | 140 792 | RM | 140 820 | EDO | 140 848 | EDO | 140 875 | FF1 |
| 140 766 | HS | 140 793 | EDO | 140 821 | EDO | 140 849 | EDO | 140 876 | FF1 |
| 140 767 | HS | 140 794 | FF1 | 140 823 | EDO | 140 850 | EDO | 140 877 | FF1 |
| 140 768 | HS | 140 795 | FF1 | 140 824 | EDO | 140 851 | EDO | 140 878 | FF1 |
| 140 769 | HS | 140 796 | FF1 | 140 825 | EDO | 140 852 | EDO | 140 879 | FF1 |
| | | 140 797 | FF1 | | | | | | |

## CLASS 141                                                       Bo-Bo

**Built:** 1956-69 by Henschel/Krauss-Maffei/Krupp.
**Electrical Equipment:** BBC/Siemens/AEG.      **Continuous Rating:** 2310 kW.
**Weight:** 66.40 tonnes.      **Maximum Speed:** 120 km/h.
**Length:** 15.620 m.      **Train Supply:** Electric.

| | | | | | | | | | |
|---|---|---|---|---|---|---|---|---|---|
| 141 001 | FF1 | 141 048 | FF1 | 141 081 | HS | 141 120 | EHG | 141 152 | HS |
| 141 003 | SSH | 141 049 | FF1 | 141 082 | FF1 | 141 121 | SSH | 141 153 | FF1 |
| 141 007 | NN1 | 141 050 | NN1 | 141 083 | HS | 141 122 | NN1 | 141 154 | SSH |
| 141 009 | NN1 | 141 051 | NN1 | 141 084 | HS | 141 123 | NN1 | 141 155 | HS |
| 141 010 | SSH | 141 052 | NN1 | 141 085 | FF1 | 141 124 | NN1 | 141 156 | NN1 |
| 141 011 | NN1 | 141 053 | NN1 | 141 086 | FF1 | 141 125 | NN1 | 141 157 | EHG |
| 141 012 | NN1 | 141 054 | RM | 141 088 | HS | 141 126 | NN1 | 141 158 | FF1 |
| 141 013 | NN1 | 141 055 | HS | 141 089 | FF1 | 141 127 | NN1 | 141 159 | EHG |
| 141 016 | NN1 | 141 056 | RM | 141 090 | HS | 141 130 | FF1 | 141 161 | HS |
| 141 017 | NN1 | 141 057 | RM | 141 091 | FF1 | 141 131 | (Z) | 141 162 | FF1 |
| 141 018 | NN1 | 141 058 | RM | 141 093 | FF1 | 141 132 | HS | 141 163 | EHG |
| 141 019 | NN1 | 141 060 | RM | 141 094 | HS | 141 133 | SSH | 141 164 | FF1 |
| 141 021 | NN1 | 141 061 | HS | 141 097 | FF1 | 141 134 | NN1 | 141 165 | (Z) |
| 141 023 | NN1 | 141 062 | NN1 | 141 098 | NN1 | 141 135 | SSH | 141 166 | HS |
| 141 024 | RM | 141 063 | SSH | 141 099 | NN1 | 141 136 | HS | 141 168 | HS |
| 141 026 | NN1 | 141 064 | FF1 | 141 100 | FF1 | 141 137 | FF1 | 141 169 | HS |
| 141 027 | NN1 | 141 065 | FF1 | 141 101 | FF1 | 141 138 | HS | 141 170 | HS |
| 141 030 | NN1 | 141 066 | FF1 | 141 102 | FF1 | 141 139 | FF1 | 141 171 | HS |
| 141 031 | NN1 | 141 067 | FF1 | 141 103 | FF1 | 141 140 | HS | 141 172 | HS |
| 141 032 | NN1 | 141 068 | FF1 | 141 106 | FF1 | 141 141 | HS | 141 173 | HS |
| 141 035 | NN1 | 141 069 | FF1 | 141 107 | FF1 | 141 142 | FF1 | 141 174 | FF1 |
| 141 036 | NN1 | 141 071 | HS | 141 108 | NN1 | 141 143 | HS | 141 175 | EHG |
| 141 037 | NN1 | 141 072 | HS | 141 110 | HS | 141 144 | FF1 | 141 176 | RM |
| 141 038 | NN1 | 141 073 | NN1 | 141 111 | SSH | 141 146 | HS | 141 177 | RM |
| 141 039 | (Z) | 141 074 | HS | 141 115 | SSH | 141 147 | FF1 | 141 178 | FF1 |
| 141 042 | SSH | 141 075 | HS | 141 116 | HS | 141 148 | NN1 | 141 179 | HS |
| 141 044 | RM | 141 078 | SSH | 141 117 | (Z) | 141 149 | FF1 | 141 180 | HS |
| 141 045 | EHG | 141 080 | HS | 141 118 | EHG | 141 151 | FF1 | 141 182 | HS |

# GERMANY

| | | | | | | | | | |
|---|---|---|---|---|---|---|---|---|---|
| 141 183 | HS | 141 242 | FF1 | 141 295 | SSH | 141 350 | HS | 141 401 | SSH |
| 141 184 | HS | 141 243 | HS | 141 296 | EHG | 141 351 | HS | 141 402 | SSH |
| 141 185 | FF1 | 141 244 | FF1 | 141 297 | EHG | 141 352 | NN1 | 141 403 | RM |
| 141 186 | FF1 | 141 245 | HS | 141 298 | HS | 141 353 | HS | 141 404 | AH1 |
| 141 187 | SSH | 141 246 | HS | 141 299 | RM | 141 354 | NN1 | 141 405 | SSH |
| 141 189 | FF1 | 141 247 | FF1 | 141 300 | EHG | 141 355 | FF1 | 141 406 | SSH |
| 141 190 | FF1 | 141 248 | EHG | 141 301 | EHG | 141 356 | NN1 | 141 407 | RM |
| 141 192 | FF1 | 141 249 | EHG | 141 302 | EHG | 141 357 | NN1 | 141 408 | SSH |
| 141 193 | HS | 141 250 | RM | 141 303 | EHG | 141 358 | NN1 | 141 410 | RM |
| 141 194 | RM | 141 251 | FF1 | 141 304 | EHG | 141 359 | NN1 | 141 411 | HS |
| 141 198 | FF1 | 141 252 | FF1 | 141 306 | EHG | 141 360 | NN1 | 141 412 | SSH |
| 141 199 | NN1 | 141 253 | RM | 141 307 | NN1 | 141 361 | HS | 141 413 | SSH |
| 141 200 | FF1 | 141 254 | HS | 141 308 | FF1 | 141 362 | NN1 | 141 414 | AH1 |
| 141 201 | HS | 141 255 | HS | 141 309 | HS | 141 363 | NN1 | 141 415 | NN1 |
| 141 202 | HS | 141 256 | HS | 141 310 | HS | 141 364 | NN1 | 141 416 | EHG |
| 141 203 | HS | 141 257 | RM | 141 311 | FF1 | 141 365 | NN1 | 141 417 | AH1 |
| 141 204 | EHG | 141 258 | EHG | 141 312 | FF1 | 141 366 | NN1 | 141 418 | NN1 |
| 141 205 | FF1 | 141 259 | FF1 | 141 313 | FF1 | 141 367 | NN1 | 141 419 | AH1 |
| 141 206 | HS | 141 260 | FF1 | 141 314 | FF1 | 141 368 | FF1 | 141 420 | HS |
| 141 207 | EHG | 141 261 | HS | 141 315 | SSH | 141 369 | NN1 | 141 421 | AH1 |
| 141 208 | HS | 141 262 | FF1 | 141 316 | FF1 | 141 370 | NN1 | 141 422 | AH1 |
| 141 210 | HS | 141 263 | HS | 141 317 | FF1 | 141 371 | NN1 | 141 423 | AH1 |
| 141 211 | HS | 141 264 | HS | 141 318 | SSH | 141 372 | NN1 | 141 424 | AH1 |
| 141 212 | RM | 141 265 | FF1 | 141 319 | FF1 | 141 373 | NN1 | 141 425 | AH1 |
| 141 213 | EHG | 141 266 | HS | 141 320 | HS | 141 374 | NN1 | 141 426 | AH1 |
| 141 214 | FF1 | 141 267 | RM | 141 321 | HS | 141 375 | NN1 | 141 427 | AH1 |
| 141 215 | RM | 141 268 | EHG | 141 322 | HS | 141 376 | NN1 | 141 428 | AH1 |
| 141 216 | RM | 141 269 | HS | 141 324 | HS | 141 377 | NN1 | 141 429 | RM |
| 141 217 | RM | 141 270 | EHG | 141 325 | HS | 141 378 | EHG | 141 430 | AH1 |
| 141 218 | NN1 | 141 271 | RM | 141 326 | EHG | 141 379 | EHG | 141 431 | HS |
| 141 219 | NN1 | 141 272 | EHG | 141 327 | EHG | 141 380 | EHG | 141 432 | HS |
| 141 221 | RM | 141 273 | (Z) | 141 328 | HS | 141 381 | NN1 | 141 433 | HS |
| 141 222 | RM | 141 274 | (Z) | 141 329 | HS | 141 382 | NN1 | 141 434 | EHG |
| 141 223 | SSH | 141 275 | EHG | 141 330 | EHG | 141 383 | SSH | 141 435 | EHG |
| 141 224 | RM | 141 276 | FF1 | 141 332 | EHG | 141 384 | NN1 | 141 436 | NN1 |
| 141 226 | FF1 | 141 277 | FF1 | 141 333 | HS | 141 385 | FF1 | 141 437 | NN1 |
| 141 227 | FF1 | 141 278 | FF1 | 141 334 | HS | 141 386 | NN1 | 141 438 | NN1 |
| 141 228 | FF1 | 141 280 | FF1 | 141 336 | HS | 141 387 | NN1 | 141 439 | NN1 |
| 141 229 | FF1 | 141 281 | HS | 141 337 | RM | 141 388 | EHG | 141 440 | NN1 |
| 141 230 | FF1 | 141 283 | HS | 141 338 | SSH | 141 389 | SSH | 141 441 | NN1 |
| 141 231 | FF1 | 141 284 | HS | 141 339 | RM | 141 390 | NN1 | 141 442 | NN1 |
| 141 232 | FF1 | 141 285 | HS | 141 340 | HS | 141 391 | SSH | 141 443 | RM |
| 141 233 | FF1 | 141 286 | EHG | 141 341 | HS | 141 392 | SSH | 141 444 | EHG |
| 141 234 | FF1 | 141 287 | EHG | 141 342 | HS | 141 393 | SSH | 141 445 | EHG |
| 141 235 | HS | 141 288 | EHG | 141 343 | HS | 141 394 | SSH | 141 446 | EHG |
| 141 236 | RM | 141 289 | EHG | 141 344 | HS | 141 395 | SSH | 141 447 | EHG |
| 141 237 | NN1 | 141 290 | EHG | 141 345 | HS | 141 396 | SSH | 141 448 | EHG |
| 141 238 | NN1 | 141 291 | FF1 | 141 346 | RM | 141 397 | NN1 | 141 449 | EHG |
| 141 239 | NN1 | 141 292 | HS | 141 347 | SSH | 141 399 | SSH | 141 450 | EHG |
| 141 240 | FF1 | 141 293 | FF1 | 141 348 | SSH | 141 400 | SSH | 141 451 | EHG |
| 141 241 | FF1 | 141 294 | EHG | 141 349 | HS | | | | |

# CLASS 142                                                      Bo-Bo

**Built:** 1963-76 by LEW.
**Electrical Equipment:** LEW.
**Weight:** 82.50 tonnes.
**Length:** 16.260 m.
**Note:** Previously DR Class 242.

**Continuous Rating:** 2740 kW.
**Maximum Speed:** 100 km/h.
**Train Supply:** Electric.

| | | | | | | | | | |
|---|---|---|---|---|---|---|---|---|---|
| 142 004 | LH2 | 142 015 | LH2 | 142 043 | LH2 | 142 063 | LW | 142 098 | LW |
| 142 007 | LH2 | 142 016 | LH2 | 142 049 | LH2 | 142 065 | LH2 | 142 102 | LH2 |
| 142 009 | LH2 | 142 019 | LH2 | 142 056 | LH2 | 142 071 | LH2 | 142 105 | LH2 |
| 142 012 | LH2 | 142 035 | LH2 | 142 059 | LH2 | 142 091 | LH2 | 142 107 | LH2 |

| | | | | | | | | | |
|---|---|---|---|---|---|---|---|---|---|
| 142 113 | LH2 | 142 175 | BSE | 142 216 | (Z) | 142 247 | LW | 142 273 | LH2 |
| 142 115 | LH2 | 142 176 | BSE | 142 217 | LH2 | 142 248 | LH2 | 142 274 | (Z) |
| 142 117 | (Z) | 142 177 | DH | 142 220 | DH | 142 249 | LW | 142 277 | (Z) |
| 142 120 | LH2 | 142 178 | LH2 | 142 222 | (Z) | 142 250 | LW | 142 280 | LH2 |
| 142 121 | BSE | 142 179 | BSE | 142 224 | (Z) | 142 252 | LW | 142 283 | DH |
| 142 123 | BSE | 142 180 | BSE | 142 226 | LH2 | 142 253 | LH2 | 142 287 | (Z) |
| 142 124 | LH2 | 142 184 | DH | 142 227 | LH2 | 142 254 | LW | 142 288 | (Z) |
| 142 136 | BSE | 142 185 | BSE | 142 228 | LH2 | 142 255 | LH2 | 142 289 | LH2 |
| 142 137 | BSE | 142 186 | (Z) | 142 229 | LH2 | 142 256 | LH2 | 142 290 | BSE |
| 142 143 | LH2 | 142 187 | DH | 142 231 | DH | 142 258 | LH2 | 142 351 | BSE |
| 142 146 | DH | 142 188 | BSE | 142 233 | BSE | 142 260 | LH2 | 142 355 | LH2 |
| 142 158 | BSE | 142 193 | (Z) | 142 238 | DH | 142 261 | LH2 | 142 366 | DH |
| 142 165 | DH | 142 195 | (Z) | 142 239 | DH | 142 264 | (Z) | 142 374 | LH2 |
| 142 166 | DZW | 142 198 | BSE | 142 242 | DH | 142 267 | (Z) | 142 376 | LH2 |
| 142 168 | BSE | 142 205 | LH2 | 142 243 | LW | 142 271 | LH2 | 142 393 | LH2 |
| 142 169 | BSE | 142 210 | LH2 | 142 244 | LH2 | 142 272 | (Z) | 142 395 | DH |
| 142 173 | DH | 142 212 | LH2 | 142 245 | LW | | | | |

## CLASS 143            Bo-Bo

**Built:** 1984-89 by LEW.
**Electrical Equipment:** LEW.
**Weight:** 82.00 tonnes.
**Length:** 16.640 m.
**Note:** Previously DR Class 243.

**Continuous Rating:** 3540 kW.
**Maximum Speed:** 120 km/h.
**Train Supply:** Electric.

| | | | | | | | | | | | |
|---|---|---|---|---|---|---|---|---|---|---|---|
| 143 002 | LH2 | 143 041 | DH | 143 080 | WR | 143 119 | LM | 143 157 | BCS |
| 143 003 | KD | 143 042 | BCS | 143 081 | DH | 143 120 | LM | 143 158 | LH2 |
| 143 005 | LL2 | 143 043 | LL2 | 143 082 | LM | 143 121 | WR | 143 159 | LM |
| 143 006 | WR | 143 044 | LL2 | 143 083 | LM | 143 122 | LM | 143 160 | WR |
| 143 007 | DH | 143 045 | KD | 143 084 | LH2 | 143 123 | LM | 143 161 | LM |
| 143 008 | DH | 143 046 | LH2 | 143 085 | DH | 143 124 | LH2 | 143 162 | WSR |
| 143 009 | DH | 143 047 | WR | 143 086 | LH2 | 143 125 | LM | 143 163 | LM |
| 143 010 | DH | 143 048 | DH | 143 087 | LL2 | 143 126 | BCS | 143 164 | WSR |
| 143 011 | LH2 | 143 049 | DH | 143 088 | WR | 143 127 | BCS | 143 165 | WR |
| 143 012 | LL2 | 143 050 | WR | 143 089 | DH | 143 128 | LM | 143 166 | WR |
| 143 013 | WSR | 143 052 | LM | 143 090 | DH | 143 129 | LM | 143 167 | LH2 |
| 143 014 | LL2 | 143 053 | DH | 143 091 | LM | 143 130 | LH2 | 143 168 | WSR |
| 143 015 | DH | 143 054 | LH2 | 143 092 | LH2 | 143 131 | LM | 143 169 | LH2 |
| 143 017 | DH | 143 055 | LH2 | 143 093 | WR | 143 132 | BCS | 143 170 | WSR |
| 143 018 | LH2 | 143 056 | LH2 | 143 094 | WR | 143 133 | LH2 | 143 171 | LH2 |
| 143 019 | LL2 | 143 057 | LL2 | 143 095 | WR | 143 134 | BCS | 143 172 | LH2 |
| 143 020 | LM | 143 058 | LL2 | 143 097 | LH2 | 143 135 | WR | 143 173 | WR |
| 143 021 | DH | 143 059 | LL2 | 143 098 | LM | 143 136 | WR | 143 174 | LH2 |
| 143 022 | DH | 143 061 | WR | 143 100 | LM | 143 137 | LH2 | 143 175 | WSR |
| 143 023 | LL2 | 143 062 | WR | 143 101 | LM | 143 138 | BCS | 143 176 | DH |
| 143 024 | DH | 143 063 | LH2 | 143 102 | DH | 143 139 | LH2 | 143 177 | KD |
| 143 025 | LM | 143 064 | LL2 | 143 103 | LH2 | 143 140 | LH2 | 143 178 | LH2 |
| 143 026 | DH | 143 065 | WR | 143 104 | LM | 143 141 | LH2 | 143 179 | WSR |
| 143 027 | LL2 | 143 066 | WR | 143 105 | LM | 143 143 | LH2 | 143 180 | LH2 |
| 143 028 | LH2 | 143 067 | WR | 143 106 | LM | 143 144 | LH2 | 143 181 | WSR |
| 143 029 | LL2 | 143 068 | LL2 | 143 107 | WR | 143 145 | WR | 143 182 | LM |
| 143 030 | KD | 143 069 | DH | 143 108 | DH | 143 146 | LM | 143 183 | WR |
| 143 031 | DH | 143 070 | LH2 | 143 109 | LM | 143 147 | WR | 143 184 | LM |
| 143 032 | LH2 | 143 071 | LH2 | 143 110 | WR | 143 148 | LH2 | 143 185 | DH |
| 143 033 | DH | 143 072 | WR | 143 111 | WR | 143 149 | DH | 143 186 | WR |
| 143 034 | LL2 | 143 073 | LH2 | 143 112 | DH | 143 150 | LM | 143 187 | LM |
| 143 035 | LM | 143 074 | LM | 143 113 | LM | 143 151 | LM | 143 188 | WSR |
| 143 036 | KD | 143 075 | WR | 143 114 | LM | 143 152 | LH2 | 143 189 | WR |
| 143 037 | LL2 | 143 076 | WR | 143 115 | LM | 143 153 | LH2 | 143 190 | LM |
| 143 038 | DH | 143 077 | LH2 | 143 116 | LW | 143 154 | LM | 143 191 | LM |
| 143 039 | WR | 143 078 | WR | 143 117 | LH2 | 143 155 | LM | 143 192 | LH2 |
| 143 040 | DH | 143 079 | DH | 143 118 | BCS | 143 156 | LM | 143 193 | WSR |

# GERMANY

| | | | | | | | | | |
|---|---|---|---|---|---|---|---|---|---|
| 143 194 | WSR | 143 260 | WSR | 143 328 | WR | 143 571 | LM | 143 636 | LM |
| 143 195 | LL2 | 143 261 | WSR | 143 329 | KD | 143 572 | KD | 143 637 | LH2 |
| 143 196 | BCS | 143 262 | BCS | 143 330 | KD | 143 573 | LL2 | 143 638 | LL2 |
| 143 197 | BCS | 143 263 | LH2 | 143 331 | LL2 | 143 574 | BCS | 143 639 | DH |
| 143 198 | LH2 | 143 265 | BCS | 143 332 | BCS | 143 575 | WR | 143 640 | WR |
| 143 199 | LM | 143 267 | WSR | 143 333 | LM | 143 576 | EDO | 143 641 | BCS |
| 143 200 | LL2 | 143 268 | DH | 143 334 | KD | 143 577 | RM | 143 642 | BCS |
| 143 201 | LM | 143 269 | WSR | 143 335 | WR | 143 578 | KD | 143 643 | KD |
| 143 202 | LM | 143 270 | WSR | 143 336 | KD | 143 579 | RM | 143 644 | LH2 |
| 143 203 | WSR | 143 271 | WSR | 143 337 | LH2 | 143 580 | LL2 | 143 645 | LH2 |
| 143 204 | WR | 143 272 | WR | 143 338 | DH | 143 581 | KD | 143 646 | LL2 |
| 143 205 | LL2 | 143 273 | LH2 | 143 339 | BCS | 143 582 | KD | 143 647 | BCS |
| 143 206 | LL2 | 143 274 | WSR | 143 340 | LL2 | 143 583 | RM | 143 648 | LL2 |
| 143 207 | WR | 143 275 | WSR | 143 341 | DH | 143 584 | KD | 143 649 | LL2 |
| 143 208 | DH | 143 276 | WSR | 143 342 | LM | 143 585 | LL2 | 143 650 | LH2 |
| 143 210 | BCS | 143 277 | WSR | 143 343 | LH2 | 143 586 | KD | 143 651 | BCS |
| 143 211 | LL2 | 143 278 | WSR | 143 344 | BCS | 143 587 | KD | 143 652 | KD |
| 143 212 | WSR | 143 279 | WSR | 143 345 | DH | 143 588 | KD | 143 653 | BCS |
| 143 213 | WSR | 143 280 | WSR | 143 346 | WR | 143 589 | LL2 | 143 654 | BCS |
| 143 214 | LM | 143 281 | WSR | 143 347 | LL2 | 143 590 | KD | 143 655 | LM |
| 143 215 | KD | 143 282 | WSR | 143 348 | LH2 | 143 591 | LL2 | 143 656 | DH |
| 143 216 | LM | 143 283 | BCS | 143 349 | LH2 | 143 592 | KD | 143 657 | BCS |
| 143 217 | DH | 143 284 | LH2 | 143 350 | WR | 143 593 | KD | 143 658 | LL2 |
| 143 218 | WR | 143 285 | WSR | 143 351 | KD | 143 594 | KD | 143 659 | NN1 |
| 143 220 | BCS | 143 286 | WSR | 143 352 | LL2 | 143 595 | LL2 | 143 660 | UE |
| 143 221 | LH2 | 143 287 | WR | 143 353 | LH2 | 143 596 | RM | 143 661 | WR |
| 143 222 | LH2 | 143 288 | KD | 143 354 | LH2 | 143 597 | RM | 143 662 | LH2 |
| 143 224 | LH2 | 143 289 | BCS | 143 355 | LH2 | 143 598 | NN1 | 143 801 | BCS |
| 143 225 | LH2 | 143 290 | LH2 | 143 356 | LM | 143 599 | KD | 143 802 | WR |
| 143 226 | LM | 143 291 | WR | 143 357 | KD | 143 600 | KD | 143 803 | DH |
| 143 227 | LH2 | 143 292 | KD | 143 358 | KD | 143 601 | UE | 143 804 | DH |
| 143 228 | WSR | 143 293 | BCS | 143 359 | DH | 143 602 | RM | 143 805 | DH |
| 143 229 | LH2 | 143 294 | BCS | 143 360 | WSR | 143 603 | UE | 143 806 | DH |
| 143 230 | DH | 143 295 | BCS | 143 361 | DH | 143 604 | KD | 143 807 | BCS |
| 143 231 | WR | 143 298 | KD | 143 362 | DH | 143 605 | KD | 143 808 | WR |
| 143 232 | WR | 143 299 | WR | 143 363 | LM | 143 606 | RM | 143 809 | LL2 |
| 143 233 | WSR | 143 300 | LL2 | 143 364 | KD | 143 607 | KD | 143 810 | BCS |
| 143 234 | LM | 143 301 | DH | 143 365 | DH | 143 608 | KD | 143 811 | DH |
| 143 235 | WR | 143 302 | LH2 | 143 366 | LM | 143 609 | LM | 143 812 | LL2 |
| 143 236 | WSR | 143 303 | LL2 | 143 367 | DH | 143 610 | LH2 | 143 813 | LL2 |
| 143 237 | KD | 143 304 | KD | 143 368 | DH | 143 611 | KD | 143 814 | LL2 |
| 143 238 | WSR | 143 305 | LH2 | 143 369 | LH2 | 143 612 | RM | 143 815 | KD |
| 143 239 | LH2 | 143 306 | LL2 | 143 370 | LH2 | 143 613 | KD | 143 816 | LH2 |
| 143 240 | WSR | 143 307 | WR | 143 551 | LL2 | 143 614 | KD | 143 817 | LH2 |
| 143 241 | WSR | 143 308 | WR | 143 552 | RM | 143 615 | KD | 143 818 | BCS |
| 143 242 | WSR | 143 309 | LL2 | 143 553 | RM | 143 616 | LH2 | 143 819 | LH2 |
| 143 243 | DH | 143 310 | DH | 143 554 | LL2 | 143 617 | RM | 143 820 | BCS |
| 143 244 | LH2 | 143 311 | LH2 | 143 555 | UE | 143 618 | KD | 143 821 | BCS |
| 143 245 | WR | 143 312 | WR | 143 556 | BCS | 143 619 | KD | 143 822 | LM |
| 143 246 | LM | 143 313 | WR | 143 557 | KD | 143 621 | NN1 | 143 823 | KD |
| 143 247 | WR | 143 314 | DH | 143 558 | LM | 143 623 | (Z) | 143 824 | DH |
| 143 248 | WSR | 143 315 | WR | 143 559 | BCS | 143 624 | NN1 | 143 825 | LH2 |
| 143 249 | BCS | 143 316 | WR | 143 560 | LL2 | 143 625 | NN1 | 143 826 | LL2 |
| 143 250 | WSR | 143 317 | KD | 143 561 | LH2 | 143 626 | NN1 | 143 827 | WR |
| 143 251 | WSR | 143 318 | BCS | 143 562 | LL2 | 143 627 | LM | 143 828 | LM |
| 143 252 | BCS | 143 319 | LH2 | 143 563 | LL2 | 143 628 | NN1 | 143 829 | LH2 |
| 143 253 | LH2 | 143 320 | WR | 143 564 | BCS | 143 629 | NN1 | 143 830 | KD |
| 143 254 | BCS | 143 321 | WR | 143 565 | EDO | 143 630 | WR | 143 831 | BCS |
| 143 255 | LH2 | 143 323 | WR | 143 566 | EDO | 143 631 | WSR | 143 832 | DH |
| 143 256 | BCS | 143 324 | WR | 143 567 | WR | 143 632 | NN1 | 143 833 | LL2 |
| 143 257 | WR | 143 325 | DH | 143 568 | WR | 143 633 | (Z) | 143 834 | KD |
| 143 258 | WR | 143 326 | BCS | 143 569 | WR | 143 634 | NN1 | 143 835 | BCS |
| 143 259 | WR | 143 327 | WR | 143 570 | WSR | 143 635 | KD | 143 836 | KD |

| | | | | | | | | | |
|---|---|---|---|---|---|---|---|---|---|
| 143 837 | BCS | 143 865 | BCS | 143 892 | EDO | 143 920 | RM | 143 947 | EDO |
| 143 838 | WR | 143 866 | BCS | 143 893 | LL2 | 143 921 | EDO | 143 948 | EDO |
| 143 839 | WR | 143 867 | DH | 143 894 | EDO | 143 922 | UE | 143 949 | EDO |
| 143 840 | KD | 143 868 | WR | 143 895 | EDO | 143 923 | RM | 143 950 | EDO |
| 143 841 | LL2 | 143 869 | BCS | 143 896 | RM | 143 924 | EDO | 143 951 | EDO |
| 143 842 | KD | 143 870 | KD | 143 897 | RM | 143 925 | RM | 143 952 | RM |
| 143 843 | BCS | 143 871 | EDO | 143 898 | EDO | 143 926 | RM | 143 953 | BCS |
| 143 844 | BCS | 143 872 | RM | 143 899 | RM | 143 927 | RM | 143 954 | RM |
| 143 845 | WR | 143 873 | RM | 143 900 | EDO | 143 928 | EDO | 143 955 | EDO |
| 143 846 | DH | 143 874 | WR | 143 901 | EDO | 143 929 | RM | 143 956 | BCS |
| 143 847 | DH | 143 875 | DH | 143 902 | RM | 143 930 | RM | 143 957 | EDO |
| 143 848 | BCS | 143 876 | EDO | 143 903 | EDO | 143 931 | BCS | 143 958 | EDO |
| 143 849 | BCS | 143 877 | RM | 143 904 | RM | 143 932 | EDO | 143 959 | EDO |
| 143 850 | KD | 143 878 | EDO | 143 905 | RM | 143 933 | EDO | 143 960 | EDO |
| 143 851 | BCS | 143 879 | EDO | 143 906 | RM | 143 934 | RM | 143 961 | EDO |
| 143 852 | BCS | 143 880 | EDO | 143 907 | EDO | 143 935 | RM | 143 962 | RM |
| 143 853 | LL2 | 143 881 | EDO | 143 908 | RM | 143 936 | EDO | 143 963 | EDO |
| 143 854 | KD | 143 882 | EDO | 143 909 | EDO | 143 937 | EDO | 143 964 | EDO |
| 143 855 | LL2 | 143 883 | RM | 143 910 | EDO | 143 938 | BCS | 143 965 | EDO |
| 143 856 | LL2 | 143 884 | EDO | 143 911 | EDO | 143 939 | EDO | 143 966 | LH2 |
| 143 857 | LL2 | 143 885 | RM | 143 913 | EDO | 143 940 | EDO | 143 967 | EDO |
| 143 858 | LL2 | 143 886 | WR | 143 914 | RM | 143 941 | BCS | 143 968 | EDO |
| 143 859 | WR | 143 887 | EDO | 143 915 | RM | 143 942 | EDO | 143 969 | BCS |
| 143 860 | WR | 143 888 | EDO | 143 916 | RM | 143 943 | EDO | 143 970 | KD |
| 143 861 | LH2 | 143 889 | EDO | 143 917 | EDO | 143 944 | EDO | 143 971 | WR |
| 143 862 | WR | 143 890 | RM | 143 918 | EDO | 143 945 | EDO | 143 972 | BCS |
| 143 863 | LL2 | 143 891 | EDO | 143 919 | EDO | 143 946 | EDO | 143 973 | DH |
| 143 864 | LL2 | | | | | | | | |

## CLASS 150      Co-Co

**Built:** 1957-73 by Krupp.
**Electrical Equipment:** AEG.
**Weight:** 126.00-128.00 tonnes.
**Length:** 19.490 m.
**Continuous Rating:** 4410 kW.
**Maximum Speed:** 100 km/h.
**Train Supply:** Electric.

| | | | | | | | | | |
|---|---|---|---|---|---|---|---|---|---|
| 150 001 | NN2 | 150 049 | EHG | 150 078 | NN2 | 150 106 | EHG | 150 135 | EHG |
| 150 006 | NN2 | 150 050 | EHG | 150 079 | NN2 | 150 107 | EHG | 150 136 | EHG |
| 150 014 | NN2 | 150 051 | EHG | 150 080 | NN2 | 150 108 | EHG | 150 137 | EHG |
| 150 019 | NN2 | 150 052 | EHG | 150 081 | NN2 | 150 109 | EHG | 150 138 | EHG |
| 150 021 | NN2 | 150 053 | EHG | 150 082 | NN2 | 150 110 | EHG | 150 139 | EHG |
| 150 023 | NN2 | 150 054 | EHG | 150 083 | NN2 | 150 111 | EHG | 150 140 | EHG |
| 150 024 | NN2 | 150 055 | NN2 | 150 084 | NN2 | 150 112 | EHG | 150 141 | EHG |
| 150 025 | NN2 | 150 056 | EHG | 150 085 | NN2 | 150 113 | NN2 | 150 142 | EHG |
| 150 026 | NN2 | 150 057 | EHG | 150 086 | NN2 | 150 114 | NN2 | 150 143 | EHG |
| 150 027 | NN2 | 150 058 | EHG | 150 087 | NN2 | 150 115 | NN2 | 150 144 | TS |
| 150 028 | NN2 | 150 059 | EHG | 150 088 | NN2 | 150 116 | NN2 | 150 145 | TS |
| 150 029 | NN2 | 150 060 | EHG | 150 089 | NN2 | 150 119 | EHG | 150 146 | TS |
| 150 030 | NN2 | 150 061 | NN2 | 150 090 | NN2 | 150 120 | EHG | 150 147 | TS |
| 150 031 | NN2 | 150 062 | NN2 | 150 091 | NN2 | 150 121 | EHG | 150 148 | TS |
| 150 032 | NN2 | 150 063 | NN2 | 150 092 | NN2 | 150 122 | EHG | 150 149 | TS |
| 150 033 | NN2 | 150 064 | NN2 | 150 093 | NN2 | 150 123 | EHG | 150 150 | TS |
| 150 034 | NN2 | 150 065 | NN2 | 150 094 | NN2 | 150 124 | EHG | 150 151 | TS |
| 150 035 | NN2 | 150 066 | NN2 | 150 095 | NN2 | 150 125 | EHG | 150 152 | TS |
| 150 037 | NN2 | 150 067 | NN2 | 150 097 | NN2 | 150 126 | EHG | 150 153 | TS |
| 150 038 | NN2 | 150 068 | NN2 | 150 098 | NN2 | 150 127 | EHG | 150 154 | TS |
| 150 040 | NN2 | 150 070 | NN2 | 150 099 | NN2 | 150 128 | EHG | 150 155 | TS |
| 150 042 | NN2 | 150 071 | NN2 | 150 100 | EHG | 150 129 | EHG | 150 156 | TS |
| 150 043 | NN2 | 150 072 | NN2 | 150 101 | EHG | 150 130 | EHG | 150 157 | TS |
| 150 044 | NN2 | 150 073 | NN2 | 150 102 | EHG | 150 131 | EHG | 150 158 | TS |
| 150 045 | EHG | 150 075 | NN2 | 150 103 | EHG | 150 132 | EHG | 150 159 | TS |
| 150 046 | EHG | 150 076 | NN2 | 150 104 | EHG | 150 133 | EHG | 150 160 | TS |
| 150 047 | EHG | 150 077 | NN2 | 150 105 | EHG | 150 134 | EHG | 150 161 | TS |

▲ "Eurosprinter" 127 001 heads an Inter City service at München Hbf on 29.01.95
(P. Wormald)

▼ 150 181 stands at Westerstetten whilst working the 1134 Geislingen - Ulm on 08.04.94
(M. Dunn)

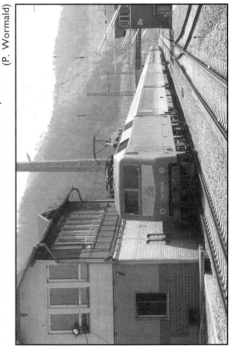

▲ 113 270 awaits departure from München with the 1700 to Mittenwald on 17.04.94
(D. Guy)

▼ 120 135 arrives at Probstzella with an Inter City service on 29.04.95
(P. Wormald)

| | | | | | | | | | |
|---|---|---|---|---|---|---|---|---|---|
| 150 162 | TS | 150 169 | TS | 150 176 | TS | 150 183 | TS | 150 189 | TS |
| 150 163 | TS | 150 170 | TS | 150 177 | TS | 150 184 | TS | 150 190 | TS |
| 150 164 | TS | 150 171 | TS | 150 178 | TS | 150 185 | TS | 150 191 | TS |
| 150 165 | TS | 150 172 | TS | 150 179 | TS | 150 186 | TS | 150 192 | TS |
| 150 166 | TS | 150 173 | TS | 150 180 | TS | 150 187 | TS | 150 193 | TS |
| 150 167 | TS | 150 174 | TS | 150 181 | TS | 150 188 | TS | 150 194 | TS |
| 150 168 | TS | 150 175 | TS | 150 182 | TS | | | | |

## CLASS 151  Co-Co

**Built:** 1973-77 by Krupp.
**Electrical Equipment:** AEG.
**Weight:** 118.00 tonnes.
**Length:** 19.490 m.
**Continuous Rating:** 5982 kW.
**Maximum Speed:** 120 km/h.
**Train Supply:** Electric.

| | | | | | | | | | |
|---|---|---|---|---|---|---|---|---|---|
| 151 001 | NN2 | 151 035 | NN2 | 151 069 | NN2 | 151 103 | NN2 | 151 137 | EHG |
| 151 002 | NN2 | 151 036 | NN2 | 151 070 | NN2 | 151 104 | NN2 | 151 138 | EHG |
| 151 003 | NN2 | 151 037 | NN2 | 151 071 | NN2 | 151 105 | EHG | 151 139 | EHG |
| 151 004 | NN2 | 151 038 | NN2 | 151 072 | NN2 | 151 106 | EHG | 151 140 | EHG |
| 151 005 | NN2 | 151 039 | NN2 | 151 073 | NN2 | 151 107 | EHG | 151 141 | EHG |
| 151 006 | NN2 | 151 040 | NN2 | 151 074 | NN2 | 151 108 | EHG | 151 142 | EHG |
| 151 007 | NN2 | 151 041 | NN2 | 151 075 | NN2 | 151 109 | EHG | 151 143 | EHG |
| 151 008 | NN2 | 151 042 | NN2 | 151 076 | NN2 | 151 110 | EHG | 151 144 | EHG |
| 151 009 | NN2 | 151 043 | NN2 | 151 077 | NN2 | 151 111 | EHG | 151 145 | EHG |
| 151 010 | NN2 | 151 044 | NN2 | 151 078 | NN2 | 151 112 | EHG | 151 146 | EHG |
| 151 011 | NN2 | 151 045 | NN2 | 151 079 | NN2 | 151 113 | EHG | 151 147 | EHG |
| 151 012 | NN2 | 151 046 | NN2 | 151 080 | NN2 | 151 114 | EHG | 151 148 | EHG |
| 151 013 | NN2 | 151 047 | NN2 | 151 081 | NN2 | 151 115 | EHG | 151 149 | EHG |
| 151 014 | NN2 | 151 048 | NN2 | 151 082 | NN2 | 151 116 | EHG | 151 150 | EHG |
| 151 015 | NN2 | 151 049 | NN2 | 151 083 | NN2 | 151 117 | EHG | 151 151 | EHG |
| 151 016 | NN2 | 151 050 | NN2 | 151 084 | NN2 | 151 118 | EHG | 151 152 | EHG |
| 151 017 | NN2 | 151 051 | NN2 | 151 085 | NN2 | 151 119 | EHG | 151 153 | EHG |
| 151 018 | NN2 | 151 052 | NN2 | 151 086 | NN2 | 151 120 | EHG | 151 154 | EHG |
| 151 019 | NN2 | 151 053 | NN2 | 151 087 | NN2 | 151 121 | EHG | 151 155 | EHG |
| 151 020 | NN2 | 151 054 | NN2 | 151 088 | NN2 | 151 122 | EHG | 151 156 | EHG |
| 151 021 | NN2 | 151 055 | NN2 | 151 089 | EHG | 151 123 | EHG | 151 157 | EHG |
| 151 022 | NN2 | 151 056 | NN2 | 151 090 | EHG | 151 124 | EHG | 151 158 | EHG |
| 151 023 | NN2 | 151 057 | NN2 | 151 091 | EHG | 151 125 | EHG | 151 159 | EHG |
| 151 024 | NN2 | 151 058 | NN2 | 151 092 | EHG | 151 126 | EHG | 151 160 | EHG |
| 151 025 | NN2 | 151 059 | NN2 | 151 093 | EHG | 151 127 | EHG | 151 161 | EHG |
| 151 026 | NN2 | 151 060 | NN2 | 151 094 | EHG | 151 128 | EHG | 151 162 | EHG |
| 151 027 | NN2 | 151 061 | NN2 | 151 095 | EHG | 151 129 | EHG | 151 163 | EHG |
| 151 028 | NN2 | 151 062 | NN2 | 151 096 | EHG | 151 130 | EHG | 151 164 | EHG |
| 151 029 | NN2 | 151 063 | NN2 | 151 097 | EHG | 151 131 | EHG | 151 165 | EHG |
| 151 030 | NN2 | 151 064 | NN2 | 151 098 | EHG | 151 132 | EHG | 151 166 | EHG |
| 151 031 | NN2 | 151 065 | NN2 | 151 099 | EHG | 151 133 | EHG | 151 167 | EHG |
| 151 032 | NN2 | 151 066 | NN2 | 151 100 | NN2 | 151 134 | EHG | 151 168 | EHG |
| 151 033 | NN2 | 151 067 | NN2 | 151 101 | NN2 | 151 135 | EHG | 151 169 | EHG |
| 151 034 | NN2 | 151 068 | NN2 | 151 102 | NN2 | 151 136 | EHG | 151 170 | EHG |

## CLASS 155  Co-Co

**Built:** 1974-84 by LEW.
**Electrical Equipment:** LEW.
**Weight:** 123.00 tonnes.
**Length:** 19.600 m.
**Note:** Previously DR Class 250.
**Continuous Rating:** 5100 kW.
**Maximum Speed:** 125 km/h.
**Train Supply:** Electric.

| | | | | | | | | | |
|---|---|---|---|---|---|---|---|---|---|
| 155 001 | LL2 | 155 008 | LL2 | 155 014 | DH | 155 020 | LL2 | 155 026 | DH |
| 155 003 | LL2 | 155 009 | LL2 | 155 015 | LL2 | 155 021 | LL2 | 155 027 | LL2 |
| 155 004 | BHW | 155 010 | RM | 155 016 | RM | 155 022 | BSE | 155 028 | BSE |
| 155 005 | DH | 155 011 | LL2 | 155 017 | LL2 | 155 023 | LL2 | 155 029 | LL2 |
| 155 006 | BSE | 155 012 | BSE | 155 018 | LL2 | 155 024 | LL2 | 155 030 | LL2 |
| 155 007 | RM | 155 013 | RM | 155 019 | RM | 155 025 | DH | 155 031 | LL2 |

# GERMANY

| | | | | | | | | | |
|---|---|---|---|---|---|---|---|---|---|
| 155 032 | LL2 | 155 080 | LL2 | 155 129 | LL2 | 155 178 | LL2 | 155 226 | LL2 |
| 155 033 | RM | 155 081 | DH | 155 130 | RM | 155 179 | LL2 | 155 227 | DH |
| 155 034 | DH | 155 082 | LL2 | 155 131 | BSE | 155 180 | RM | 155 228 | LL2 |
| 155 035 | BSE | 155 083 | BHW | 155 132 | BSE | 155 181 | BSE | 155 229 | DH |
| 155 036 | LL2 | 155 084 | BSE | 155 133 | BSE | 155 182 | DH | 155 230 | BHW |
| 155 037 | DH | 155 085 | BHW | 155 134 | BSE | 155 183 | RM | 155 231 | BHW |
| 155 038 | LL2 | 155 086 | LL2 | 155 135 | RM | 155 184 | LL2 | 155 232 | LL2 |
| 155 039 | DH | 155 087 | DH | 155 136 | RM | 155 185 | LL2 | 155 233 | BSE |
| 155 040 | LL2 | 155 088 | BSE | 155 137 | BHW | 155 186 | RM | 155 234 | BHW |
| 155 041 | LL2 | 155 089 | RM | 155 138 | BSE | 155 187 | RM | 155 235 | LL2 |
| 155 042 | LL2 | 155 090 | BHW | 155 139 | RM | 155 188 | BHW | 155 236 | LL2 |
| 155 043 | LL2 | 155 091 | BSE | 155 140 | BSE | 155 189 | LL2 | 155 237 | RM |
| 155 044 | LL2 | 155 092 | LL2 | 155 141 | BHW | 155 190 | DH | 155 238 | LL2 |
| 155 045 | LL2 | 155 093 | RM | 155 142 | DH | 155 191 | DH | 155 239 | LL2 |
| 155 046 | LL2 | 155 094 | RM | 155 143 | DH | 155 192 | BSE | 155 240 | BHW |
| 155 047 | LL2 | 155 095 | RM | 155 144 | BHW | 155 193 | DH | 155 241 | LL2 |
| 155 048 | DH | 155 096 | LL2 | 155 146 | BSE | 155 194 | LL2 | 155 242 | LL2 |
| 155 049 | DH | 155 097 | BSE | 155 147 | BSE | 155 195 | RM | 155 243 | NN2 |
| 155 050 | DH | 155 098 | BSE | 155 148 | RM | 155 196 | BSE | 155 244 | LL2 |
| 155 051 | BHW | 155 099 | BHW | 155 149 | BSE | 155 197 | NN2 | 155 245 | LL2 |
| 155 052 | RM | 155 100 | LL2 | 155 150 | DH | 155 198 | LL2 | 155 246 | BSE |
| 155 053 | LL2 | 155 101 | LL2 | 155 151 | BSE | 155 199 | LL2 | 155 247 | BSE |
| 155 054 | DH | 155 102 | BHW | 155 152 | LL2 | 155 200 | DH | 155 248 | LL2 |
| 155 055 | LL2 | 155 103 | BSE | 155 153 | DH | 155 201 | BSE | 155 249 | LL2 |
| 155 056 | BHW | 155 104 | LL2 | 155 154 | DH | 155 202 | DH | 155 250 | RM |
| 155 057 | LL2 | 155 105 | LL2 | 155 155 | BSE | 155 203 | LL2 | 155 251 | LL2 |
| 155 058 | LL2 | 155 107 | DH | 155 156 | LL2 | 155 204 | LL2 | 155 252 | RM |
| 155 059 | DH | 155 108 | LL2 | 155 157 | LL2 | 155 205 | LL2 | 155 253 | BSE |
| 155 060 | RM | 155 109 | BSE | 155 158 | DH | 155 206 | BSE | 155 254 | RM |
| 155 061 | RM | 155 110 | BSE | 155 159 | BSE | 155 207 | RM | 155 255 | BHW |
| 155 062 | NN2 | 155 111 | RM | 155 160 | BSE | 155 208 | DH | 155 256 | LL2 |
| 155 063 | BSE | 155 112 | DH | 155 161 | LL2 | 155 209 | BSE | 155 257 | LL3 |
| 155 064 | DH | 155 113 | RM | 155 162 | BSE | 155 210 | RM | 155 258 | LL2 |
| 155 065 | LL2 | 155 114 | BSE | 155 163 | LL2 | 155 211 | LL2 | 155 259 | LL2 |
| 155 066 | LL2 | 155 115 | BSE | 155 164 | LL2 | 155 212 | LL2 | 155 260 | LL2 |
| 155 067 | BSE | 155 116 | DH | 155 165 | BSE | 155 213 | BSE | 155 261 | LL2 |
| 155 068 | LL2 | 155 117 | DH | 155 166 | BSE | 155 214 | DH | 155 262 | LL2 |
| 155 069 | BHW | 155 118 | LL2 | 155 167 | BSE | 155 215 | RM | 155 263 | LL2 |
| 155 070 | LL2 | 155 119 | RM | 155 168 | RM | 155 216 | LL2 | 155 264 | RM |
| 155 071 | LL2 | 155 120 | LL2 | 155 169 | DH | 155 217 | DH | 155 265 | RM |
| 155 072 | DH | 155 121 | LL2 | 155 170 | BSE | 155 218 | LL2 | 155 266 | LL2 |
| 155 073 | BSE | 155 122 | LL2 | 155 171 | BSE | 155 219 | LL2 | 155 267 | NN2 |
| 155 074 | BSE | 155 123 | BHW | 155 172 | BSE | 155 220 | LL2 | 155 268 | LL2 |
| 155 075 | DH | 155 124 | LL2 | 155 173 | DH | 155 221 | LL2 | 155 269 | BSE |
| 155 076 | LL2 | 155 125 | LL2 | 155 174 | BSE | 155 222 | DH | 155 270 | LL2 |
| 155 077 | RM2 | 155 126 | LL2 | 155 175 | RM | 155 223 | LL2 | 155 271 | LL2 |
| 155 078 | LL2 | 155 127 | DH | 155 176 | LL2 | 155 224 | DH | 155 272 | BSE |
| 155 079 | DH | 155 128 | BSE | 155 177 | LL2 | 155 225 | DH | 155 273 | RM |

# CLASS 156                                                          Co-Co

**Built:** 1991-92 by LEW.
**Electrical Equipment:** LEW.
**Weight:** 120.00 tonnes.
**Length:** 19.500 m.
**Note:** Previously DR Class 252.

**Continuous Rating:** 5880 kW.
**Maximum Speed:** 120 km/h.
**Train Supply:** Electric.

156 001 DH     156 002 DH     156 003 DH     156 004 DH

# CLASS 171                                                          Co-Co

**Built:** 1964-65 by LEW.
**Supply System:** 25 kV 50 Hz overhead.

**Electrical Equipment:** LEW.
**Weight:** 126.00 tonnes.
**Length:** 18.640 m.
**Note:** Previously DR Class 251.

**Continuous Rating:** 3300 kW.
**Maximum Speed:** 80 km/h.
**Train Supply:** Electric.

| | | | | |
|---|---|---|---|---|
| 171 001 LHB | 171 004 LHB | 171 008 LHB | 171 011 (Z) | 171 013 LHB |
| 171 002 LHB | 171 005 LHB | 171 009 LHB | 171 012 LHB | 171 014 LHB |
| 171 003 LHB | | | | |

## CLASS 180      Bo-Bo

**Built:** 1987-91 by Skoda.
**Supply System:** 15 kV 16.67 Hz overhead or 3 kV dc overhead.
**Electrical Equipment:** Skoda.
**Weight:** 84.00 tonnes.
**Length:** 16.800 m.
**Note:** Previously DR Class 230.

**Continuous Rating:**
**Maximum Speed:** 120 km/h.
**Train Supply:** Electric.

| | | | | |
|---|---|---|---|---|
| 180 001 DH | 180 005 DH | 180 009 DH | 180 013 DH | 180 017 DH |
| 180 002 DH | 180 006 DH | 180 010 DH | 180 014 DH | 180 018 DH |
| 180 003 DH | 180 007 DH | 180 011 DH | 180 015 DH | 180 019 DH |
| 180 004 DH | 180 008 DH | 180 012 DH | 180 016 DH | 180 020 DH |

## CLASS 181      Bo-Bo

**Built:** 1967* or 1974-75 by Krupp.
**Supply System:** 15 kV ac 16.67 Hz ac overhead or 25 Kv ac 50 Hz overhead.
**Electrical Equipment:** AEG.
**Weight:** 82.50 (* 84.00) tonnes.
**Length:** 17.940 (* 16.950) m.

**Continuous Rating:** 3200 (* 3000) kW.
**Maximum Speed:** 160 (* 150) km/h.
**Train Supply:** Electric.

| | | | | |
|---|---|---|---|---|
| 181 001*SSH | 181 206 SSH | 181 211 SSH | 181 216 SSH | 181 221 SSH |
| 181 201 SSH | 181 207 SSH | 181 212 SSH | 181 217 SSH | 181 222 SSH |
| 181 202 SSH | 181 208 SSH | 181 213 SSH | 181 218 SSH | 181 223 SSH |
| 181 203 SSH | 181 209 SSH | 181 214 SSH | 181 219 SSH | 181 224 SSH |
| 181 204 SSH | 181 210 SSH | 181 215 SSH | 181 220 SSH | 181 225 SSH |
| 181 205 SSH | | | | |

**Names:** 181 211 *Lorraine*      181 213 *Saar*
        181 212 *Luxembourg*      181 214 *Mosel*

## CLASS 184      Bo-Bo

**Built:** 1966 by Krupp.
**Supply System:** 15 kV ac 16.67 Hz ac overhead or 25 Kv ac 50 Hz overhead.
**Electrical Equipment:** AEG.
**Weight:** 84.00 tonnes.
**Length:** 16.950 m.

**Continuous Rating:** 3000 kW.
**Maximum Speed:** 150 km/h.
**Train Supply:** Electric.

184 003 SSH

## DIESEL MAIN LINE LOCOMOTIVES

## CLASS 201      B-B

**Built:** 1964-78 by LEW.
**Engine:** Johannisthal 12 KVD 21 AL3 of 736 kW at 1500 rpm.
**Transmission:** Hydraulic. Strömungsmachinen GSR 30/5.7 AQ.
**Weight:** 60.00-63.70 tonnes.
**Length:** 13.940 m.
**Note:** Previously DR Class 110.

**Maximum Speed:** 100 km/h.
**Train Supply:** Steam.

| | | | | | | | | | |
|---|---|---|---|---|---|---|---|---|---|
| 201 002 | (Z) | 201 035 | (Z) | 201 038 | WNT | 201 053 | UEI | 201 070 | UW |
| 201 024 | (Z) | 201 036 | LH1 | 201 040 | (Z) | 201 059 | UW | 201 073 | (Z) |
| 201 025 | (Z) | 201 037 | LW | 201 043 | (Z) | 201 064 | (Z) | 201 076 | UW |

# GERMANY

| | | | | | | | | | |
|---|---|---|---|---|---|---|---|---|---|
| 201 082 | UW | 201 221 | (Z) | 201 493 | (Z) | 201 704 | WNT | 201 829 | (Z) |
| 201 090 | (Z) | 201 222 | UW | 201 510 | WW | 201 711 | (Z) | 201 830 | LH1 |
| 201 096 | (Z) | 201 225 | LW | 201 557 | (Z) | 201 734 | (Z) | 201 839 | UG |
| 201 101 | (Z) | 201 233 | (Z) | 201 566 | (Z) | 201 740 | (Z) | 201 851 | DRC |
| 201 118 | DRC | 201 239 | (Z) | 201 570 | LH1 | 201 745 | (Z) | 201 867 | LH1 |
| 201 123 | WNT | 201 243 | (Z) | 201 577 | UEI | 201 755 | (Z) | 201 868 | DRC |
| 201 125 | WNT | 201 259 | (Z) | 201 604 | (Z) | 201 756 | (Z) | 201 873 | (Z) |
| 201 132 | DRC | 201 273 | DRC | 201 608 | (Z) | 201 763 | (Z) | 201 875 | DRC |
| 201 133 | LW | 201 284 | (Z) | 201 609 | (Z) | 201 767 | (Z) | 201 876 | (Z) |
| 201 143 | WR | 201 285 | (Z) | 201 618 | (Z) | 201 770 | (Z) | 201 881 | (Z) |
| 201 154 | (Z) | 201 304 | (Z) | 201 619 | LH1 | 201 802 | (Z) | 201 884 | (Z) |
| 201 158 | WNT | 201 339 | (Z) | 201 631 | (Z) | 201 807 | WNT | 201 886 | (Z) |
| 201 165 | DRC | 201 343 | (Z) | 201 636 | (Z) | 201 812 | LHB | 201 887 | (Z) |
| 201 171 | UW | 201 380 | WSR | 201 644 | WNT | 201 813 | (Z) | 201 888 | (Z) |
| 201 208 | (Z) | 201 391 | (Z) | 201 653 | (Z) | 201 815 | WNT | 201 889 | (Z) |
| 201 213 | (Z) | 201 393 | UW | 201 665 | (Z) | 201 826 | LH1 | 201 895 | DRC |
| 201 215 | WNT | 201 419 | LW | 201 669 | WNT | | | | |

# CLASS 202                                                                 B-B

**Built:** 1964-78 by LEW as class 110. Rebuilt 1981 onwards by DR Stendal Works.
**Engine:** Johannisthal 12 KVD 21 AL4 of 883 kW at 1500 rpm.
**Transmission:** Hydraulic. Strömungsmachinen GSR 30/5.7 AQ.
**Weight:** 64.00 tonnes.                    **Maximum Speed:** 100 km/h.
**Length:** 13.940-14.240 m.                  **Train Supply:** Steam.
**Note:** Previously DR Class 112.

| | | | | | | | | | |
|---|---|---|---|---|---|---|---|---|---|
| 202 057 | LS | 202 242 | LHB | 202 295 | BSE | 202 338 | WNT | 202 383 | DC |
| 202 063 | DZW | 202 245 | DR | 202 296 | WNT | 202 340 | UN | 202 384 | UM |
| 202 066 | BCS | 202 248 | LM | 202 297 | LHB | 202 341 | LHB | 202 385 | UM |
| 202 078 | DG | 202 249 | LW | 202 299 | LMR | 202 342 | DC | 202 386 | BPKR |
| 202 098 | DZW | 202 250 | DZW | 202 300 | DC | 202 344 | UE | 202 387 | BHW |
| 202 105 | LH1 | 202 251 | UE | 202 301 | UM | 202 345 | WR | 202 388 | BPKR |
| 202 109 | LS | 202 253 | LHB | 202 302 | UE | 202 346 | DC | 202 389 | DG |
| 202 114 | LS | 202 254 | BFG | 202 303 | LL1 | 202 347 | UM | 202 390 | DG |
| 202 115 | LM | 202 255 | BSE | 202 305 | BPKR | 202 348 | UG | 202 392 | BPKR |
| 202 127 | LS | 202 256 | UE | 202 307 | DC | 202 349 | DG | 202 394 | BFG |
| 202 136 | WR | 202 260 | LHB | 202 309 | UM | 202 350 | LHB | 202 395 | LHB |
| 202 137 | WR | 202 262 | LL1 | 202 310 | LH1 | 202 351 | DG | 202 396 | BHW |
| 202 144 | DC | 202 263 | LW | 202 311 | DC | 202 352 | DRC | 202 397 | UE |
| 202 152 | DC | 202 264 | BHW | 202 312 | LHB | 202 353 | UE | 202 398 | DG |
| 202 160 | LM | 202 265 | BSE | 202 313 | LL1 | 202 354 | LM | 202 400 | LL1 |
| 202 162 | DC | 202 266 | LS | 202 315 | DZW | 202 355 | LH1 | 202 403 | LMR |
| 202 164 | LH1 | 202 267 | BCS | 202 316 | WNT | 202 357 | LHB | 202 405 | DZW |
| 202 166 | DRC | 202 268 | DR | 202 317 | LL1 | 202 360 | WP | 202 406 | LHB |
| 202 167 | DZW | 202 269 | BPKR | 202 318 | UE | 202 361 | LHB | 202 407 | LL1 |
| 202 169 | BSE | 202 270 | LS | 202 319 | UM | 202 362 | WW | 202 408 | WSR |
| 202 201 | BSE | 202 271 | BHW | 202 320 | BPKR | 202 363 | LHB | 202 409 | UE |
| 202 205 | BSE | 202 272 | WNT | 202 321 | UE | 202 364 | LHB | 202 411 | LMR |
| 202 207 | BHW | 202 275 | WSR | 202 322 | BSE | 202 365 | WP | 202 412 | UM |
| 202 209 | UM | 202 276 | WP | 202 323 | DZW | 202 368 | DRC | 202 413 | LHB |
| 202 214 | BHW | 202 277 | DC | 202 324 | DG | 202 369 | DG | 202 414 | WNT |
| 202 218 | UE | 202 278 | LHB | 202 325 | WP | 202 370 | BCS | 202 415 | LW |
| 202 219 | UE | 202 279 | LHB | 202 326 | WP | 202 371 | LL1 | 202 416 | DZW |
| 202 220 | LHB | 202 280 | LH1 | 202 327 | UM | 202 372 | UE | 202 417 | BSE |
| 202 231 | DZW | 202 286 | UE | 202 329 | WR | 202 373 | UE | 202 420 | WR |
| 202 232 | UE | 202 287 | BFG | 202 330 | BSE | 202 374 | DR | 202 421 | LMR |
| 202 234 | LHB | 202 288 | DC | 202 331 | DRC | 202 375 | LMR | 202 422 | UM |
| 202 235 | BHW | 202 289 | WR | 202 332 | BPKR | 202 376 | LHB | 202 423 | DG |
| 202 236 | BFG | 202 290 | DC | 202 333 | WSR | 202 377 | LHB | 202 425 | LH1 |
| 202 237 | UE | 202 291 | UM | 202 334 | UM | 202 378 | BSE | 202 426 | LH1 |
| 202 238 | LHB | 202 292 | UM | 202 335 | UM | 202 379 | UE | 202 427 | BCS |
| 202 240 | DRC | 202 293 | BSE | 202 336 | LH1 | 202 381 | LW | 202 428 | UE |
| 202 241 | LMR | 202 294 | UM | 202 337 | LHB | 202 382 | BPKR | 202 429 | DC |

| | | | | | | | | | |
|---|---|---|---|---|---|---|---|---|---|
| 202 430 | DC | 202 505 | LHB | 202 580 | WP | 202 676 | UE | 202 766 | DR |
| 202 431 | WNT | 202 506 | LHB | 202 581 | DC | 202 677 | DC | 202 768 | WS |
| 202 432 | UE | 202 507 | BCS | 202 582 | LHB | 202 678 | LHB | 202 771 | DC |
| 202 433 | DG | 202 509 | WS | 202 585 | DC | 202 679 | LL1 | 202 772 | DRC |
| 202 434 | LL1 | 202 511 | WNT | 202 586 | DC | 202 681 | DC | 202 775 | DG |
| 202 435 | UE | 202 515 | UE | 202 587 | DZW | 202 682 | LM | 202 776 | LMR |
| 202 436 | DZW | 202 516 | DC | 202 589 | DC | 202 683 | LM | 202 777 | DG |
| 202 437 | WS | 202 517 | BPKR | 202 590 | WS | 202 684 | LM | 202 778 | DC |
| 202 438 | LS | 202 518 | BPKR | 202 591 | DC | 202 685 | UM | 202 780 | BCS |
| 202 439 | UE | 202 519 | BFG | 202 593 | DZW | 202 687 | DC | 202 781 | DZW |
| 202 441 | UM | 202 520 | BPKR | 202 594 | LL1 | 202 689 | DG | 202 783 | BCS |
| 202 442 | DG | 202 521 | BPKR | 202 595 | BSE | 202 690 | DZW | 202 784 | LMR |
| 202 443 | LS | 202 522 | LH1 | 202 596 | LL1 | 202 691 | BFG | 202 785 | LMR |
| 202 444 | WS | 202 523 | WNT | 202 597 | BSE | 202 693 | DRC | 202 786 | DC |
| 202 446 | LH1 | 202 524 | WR | 202 598 | BSE | 202 695 | LMR | 202 787 | BHW |
| 202 447 | BPKR | 202 525 | BHW | 202 599 | BSE | 202 696 | LM | 202 788 | LMR |
| 202 449 | WS | 202 526 | LHB | 202 600 | BFG | 202 697 | WS | 202 791 | DC |
| 202 450 | LHB | 202 527 | UM | 202 601 | DC | 202 700 | BCS | 202 793 | (Z) |
| 202 451 | DZW | 202 528 | DR | 202 602 | UM | 202 701 | LM | 202 794 | UM |
| 202 452 | DZW | 202 529 | LM | 202 603 | UE | 202 702 | DZW | 202 795 | UM |
| 202 453 | LHB | 202 530 | BHW | 202 606 | WW | 202 703 | DG | 202 796 | DC |
| 202 454 | LL1 | 202 531 | UE | 202 610 | LMR | 202 705 | BPKR | 202 797 | WP |
| 202 455 | (Z) | 202 532 | WP | 202 611 | DC | 202 706 | (Z) | 202 799 | LL1 |
| 202 456 | LH1 | 202 533 | LS | 202 613 | UE | 202 707 | DG | 202 800 | WR |
| 202 457 | LL1 | 202 534 | LHB | 202 614 | (Z) | 202 708 | BCS | 202 804 | DG |
| 202 458 | LHB | 202 535 | LH1 | 202 615 | DR | 202 713 | LS | 202 806 | DRC |
| 202 459 | DG | 202 536 | LM | 202 617 | DRC | 202 714 | BHW | 202 808 | WNT |
| 202 460 | WW | 202 537 | LHB | 202 620 | DC | 202 715 | LM | 202 809 | WSR |
| 202 461 | WP | 202 538 | UE | 202 621 | DR | 202 716 | LM | 202 810 | LM |
| 202 462 | WW | 202 539 | LS | 202 623 | DC | 202 718 | DZW | 202 811 | DZW |
| 202 465 | BSE | 202 542 | LS | 202 624 | DC | 202 719 | DZW | 202 814 | LM |
| 202 466 | UE | 202 543 | UE | 202 625 | DC | 202 720 | DR | 202 816 | DR |
| 202 467 | LHB | 202 544 | LL1 | 202 627 | WSR | 202 721 | UE | 202 817 | WS |
| 202 470 | LW | 202 545 | WP | 202 629 | WR | 202 722 | WR | 202 818 | LM |
| 202 471 | LHB | 202 546 | LL1 | 202 630 | WW | 202 724 | BCS | 202 819 | LM |
| 202 472 | LMR | 202 547 | WNT | 202 632 | DZW | 202 725 | LH1 | 202 822 | LM |
| 202 473 | BCS | 202 548 | LS | 202 633 | DZW | 202 726 | WW | 202 824 | BHW |
| 202 474 | BPKR | 202 549 | BSE | 202 635 | DR | 202 727 | DC | 202 825 | LL1 |
| 202 475 | WSR | 202 550 | DR | 202 637 | LMR | 202 728 | DG | 202 827 | DG |
| 202 476 | WS | 202 553 | LH1 | 202 640 | LW | 202 729 | WS | 202 831 | UM |
| 202 477 | UE | 202 554 | LH1 | 202 642 | LHB | 202 730 | LM | 202 832 | LMR |
| 202 478 | LHB | 202 555 | UE | 202 643 | LW | 202 731 | BHW | 202 835 | DZW |
| 202 479 | DG | 202 556 | DC | 202 645 | LW | 202 732 | LMR | 202 836 | DRC |
| 202 480 | WS | 202 558 | LL1 | 202 646 | LW | 202 733 | BCS | 202 841 | LM |
| 202 481 | LHB | 202 559 | DC | 202 647 | DC | 202 735 | DC | 202 842 | (Z) |
| 202 483 | WP | 202 560 | DZW | 202 648 | UE | 202 736 | LM | 202 843 | DG |
| 202 484 | UE | 202 561 | DR | 202 651 | WSR | 202 737 | WSR | 202 844 | DZW |
| 202 486 | LH1 | 202 562 | DZW | 202 652 | LM | 202 738 | UE | 202 846 | DG |
| 202 487 | UE | 202 563 | DG | 202 654 | UE | 202 741 | LW | 202 847 | LM |
| 202 488 | BCS | 202 564 | DC | 202 655 | LH1 | 202 743 | DRC | 202 849 | DRC |
| 202 489 | WS | 202 565 | UE | 202 656 | DRC | 202 744 | DR | 202 850 | DC |
| 202 490 | LL1 | 202 567 | LL1 | 202 657 | BHW | 202 746 | BPKR | 202 852 | BPKR |
| 202 494 | BSE | 202 568 | DG | 202 658 | BCS | 202 747 | DG | 202 853 | BPKR |
| 202 495 | BFG | 202 569 | WNT | 202 661 | DC | 202 748 | BFG | 202 854 | LM |
| 202 496 | DZW | 202 571 | DC | 202 662 | UE | 202 749 | LW | 202 855 | LS |
| 202 497 | UE | 202 572 | LL1 | 202 663 | DZW | 202 750 | WR | 202 859 | LL1 |
| 202 498 | DR | 202 573 | BSE | 202 666 | WNT | 202 751 | BPKR | 202 866 | BHW |
| 202 499 | DC | 202 574 | DZW | 202 667 | DZW | 202 752 | LW | 202 880 | DZW |
| 202 500 | DG | 202 575 | UE | 202 670 | LW | 202 753 | LW | 202 882 | LM |
| 202 501 | WW | 202 576 | DZW | 202 672 | DR | 202 754 | LW | 202 885 | WS |
| 202 502 | DC | 202 578 | DC | 202 674 | LS | 202 764 | DC | 202 893 | BFG |
| 202 503 | DR | 202 579 | DR | 202 675 | LM | 202 765 | DG | 202 894 | LM |
| 202 504 | DC | | | | | | | | |

## CLASS 204                                            B-B

**Built:** 1964-78 by LEW as class 110. Rebuilt 1978 onwards by DR Stendal Works.
**Engine:** Johannisthal 12 KVD 21 AL4 or 12 KVD 21 AL5 of 1104 kW at 1500 rpm.
**Transmission:** Hydraulic. Strömungsmachinen GSR 30/5.7 AQ.
**Weight:** 64.00 tonnes.                  **Maximum Speed:** 100 km/h.
**Length:** 14.240 m.                     **Train Supply:** Steam.
**Note:** Previously DR Class 114.

| 204 203 | LW | 204 399 | LH1 | 204 607 | US | 204 680 | US | 204 790 | US |
|---------|-----|---------|------|---------|-----|---------|-----|---------|-----|
| 204 223 | DH | 204 401 | US | 204 612 | US | 204 686 | DH | 204 803 | US |
| 204 246 | LH1 | 204 445 | LW | 204 616 | DH | 204 693 | DC | 204 805 | DH |
| 204 252 | DH | 204 448 | DH | 204 622 | UN | 204 698 | DH | 204 820 | US |
| 204 257 | US | 204 464 | LH1 | 204 626 | DH | 204 710 | DH | 204 834 | US |
| 204 261 | DH | 204 469 | LH1 | 204 638 | US | 204 712 | US | 204 838 | US |
| 204 274 | DH | 204 482 | DH | 204 639 | US | 204 723 | DH | 204 845 | US |
| 204 282 | DH | 204 485 | DH | 204 641 | DH | 204 758 | US | 204 848 | US |
| 204 298 | DH | 204 492 | US | 204 650 | US | 204 760 | LW | 204 857 | US |
| 204 314 | US | 204 513 | LW | 204 660 | US | 204 761 | US | 204 858 | DH |
| 204 328 | LW | 204 584 | US | 204 664 | US | 204 769 | US | 204 860 | US |
| 204 358 | LW | 204 592 | US | 204 671 | DH | 204 774 | US | 204 862 | DH |
| 204 359 | DH | 204 605 | LW | 204 673 | LW | 204 789 | US | 204 869 | US |
| 204 366 | LH1 | | | | | | | | |

## CLASS 211                                            B-B

**Built:** 1961-63 by Henschel/Jung/Deutz/Krauss-Maffei/Krupp/MaK/Esslingen.
**Engine:** Mercedes 12 V 652 TZ or Maybach 12 V 538 TA or Mercedes 12 V 493 TA of 993 kW at 1500 rpm.
**Transmission:** Hydraulic. Voith L216rs.
**Weight:** 62.00 tonnes.                  **Maximum Speed:** 100 km/h.
**Length:** 12.100 m.                     **Train Supply:** Steam.

| 211 009 | HO | 211 035 | NHO | 211 052 | (Z) | 211 141 | (Z) | 211 279 | (Z) |
|---------|-----|---------|------|---------|------|---------|------|---------|------|
| 211 011 | HO | 211 036 | NHO | 211 054 | NHO | 211 161 | (Z) | 211 283 | NHO |
| 211 012 | HO | 211 037 | NHO | 211 055 | NHO | 211 163 | HO | 211 290 | (Z) |
| 211 015 | HO | 211 038 | NHO | 211 056 | NHO | 211 195 | (Z) | 211 301 | (Z) |
| 211 019 | NHO | 211 039 | NHO | 211 057 | NHO | 211 200 | HO | 211 302 | NHO |
| 211 020 | NHO | 211 040 | NHO | 211 058 | NHO | 211 204 | HO | 211 307 | NHO |
| 211 023 | NHO | 211 041 | NHO | 211 059 | NHO | 211 205 | (Z) | 211 313 | (Z) |
| 211 024 | NHO | 211 042 | NHO | 211 060 | NHO | 211 206 | (Z) | 211 316 | NHO |
| 211 028 | (Z) | 211 043 | NHO | 211 061 | NHO | 211 208 | (Z) | 211 317 | NHO |
| 211 029 | NHO | 211 044 | NHO | 211 062 | NHO | 211 210 | (Z) | 211 319 | (Z) |
| 211 030 | NHO | 211 045 | NHO | 211 063 | NHO | 211 218 | (Z) | 211 321 | NHO |
| 211 031 | NHO | 211 049 | NHO | 211 098 | HO | 211 244 | HO | 211 332 | (Z) |
| 211 032 | NHO | 211 050 | (Z) | 211 101 | (Z) | 211 262 | (Z) | 211 344 | NHO |
| 211 033 | NHO | 211 051 | NHO | 211 123 | (Z) | 211 277 | (Z) | 211 363 | (Z) |
| 211 034 | NHO | | | | | | | | |

## CLASS 212                                            B-B

**Built:** 1958-66 by MaK/Henschel/Jung/Deutz/Krauss-Maffei/Krupp/Esslingen.
**Engine:** Mercedes 12 V 652 TZ of 993 kW at 1500 rpm.
**Transmission:** Hydraulic. Voith L216rs.
**Weight:** 63.00 tonnes.                  **Maximum Speed:** 100 km/h.
**Length:** 12.300 (* 12.100) m.           **Train Supply:** Steam.

| 212 001* | FD | 212 010* | HBS | 212 019* | FD | 212 027 | FD | 212 037 | MMF |
|----------|-----|----------|------|----------|-----|---------|-----|---------|------|
| 212 002* | (Z) | 212 011* | HG | 212 020* | MKP | 212 029 | FD | 212 038 | FD |
| 212 003* | AL | 212 012* | HBS | 212 021* | AL | 212 030 | FD | 212 039 | MMF |
| 212 004* | HBS | 212 013* | AL | 212 022 | (Z) | 212 031 | MKP | 212 041 | TK |
| 212 005* | AL | 212 014* | FG | 212 023 | FD | 212 032 | MMF | 212 042 | TK |
| 212 007* | AL | 212 015* | MKP | 212 024 | FD | 212 034 | NWH | 212 043 | MMF |
| 212 008* | HBS | 212 017* | MMF | 212 025 | FD | 212 035 | NWH | 212 044 | MMF |
| 212 009* | HBS | 212 018* | MKP | 212 026 | HG | 212 036 | MMF | 212 045 | HG |

| | | | | | | | | | |
|---|---|---|---|---|---|---|---|---|---|
| 212 047 | TK | 212 108 | NHO | 212 183 | TK | 212 250 | HG | 212 309 | EHG |
| 212 049 | FD | 212 109 | (Z) | 212 184 | TK | 212 252 | NWH | 212 310 | AL |
| 212 050 | KM | 212 111 | MMF | 212 185 | TK | 212 253 | KK | 212 311 | NWH |
| 212 051 | FD | 212 112 | FG | 212 186 | TK | 212 254 | NWH | 212 312 | EHG |
| 212 052 | FD | 212 113 | (Z) | 212 187 | TK | 212 255 | KK | 212 313 | EHG |
| 212 053 | FD | 212 116 | FG | 212 188 | AL | 212 256 | HG | 212 314 | EHG |
| 212 054 | FD | 212 117 | HG | 212 189 | (Z) | 212 258 | TK | 212 315 | EHG |
| 212 055 | HBS | 212 119 | FG | 212 190 | AL | 212 259 | NWH | 212 316 | (Z) |
| 212 056 | FD | 212 121 | FG | 212 194 | NWH | 212 261 | SSH | 212 317 | KK |
| 212 057 | KM | 212 123 | RK | 212 195 | FD | 212 262 | KM | 212 318 | AL |
| 212 058 | FD | 212 125 | FG | 212 196 | (Z) | 212 263 | HO | 212 319 | (Z) |
| 212 059 | FD | 212 126 | FG | 212 197 | FD | 212 264 | NWH | 212 320 | EHG |
| 212 060 | NWH | 212 127 | FG | 212 198 | FD | 212 265 | KM | 212 321 | KK |
| 212 061 | NWH | 212 128 | MKP | 212 200 | TK | 212 266 | KK | 212 322 | EHG |
| 212 062 | FD | 212 130 | MMF | 212 201 | TK | 212 267 | TK | 212 323 | FG |
| 212 063 | NWH | 212 131 | AL | 212 202 | HG | 212 268 | (Z) | 212 324 | KK |
| 212 064 | NWH | 212 132 | FG | 212 203 | FG | 212 269 | (Z) | 212 325 | FG |
| 212 065 | NWH | 212 133 | MMF | 212 204 | FG | 212 270 | KK | 212 326 | NWH |
| 212 066 | TK | 212 136 | TK | 212 205 | HG | 212 272 | HO | 212 327 | KK |
| 212 068 | NWH | 212 137 | TK | 212 208 | MMF | 212 273 | HO | 212 328 | KK |
| 212 069 | NWH | 212 139 | TK | 212 209 | MMF | 212 274 | NWH | 212 329 | NWH |
| 212 070 | RK | 212 146 | MKP | 212 210 | RK | 212 275 | AL | 212 330 | NWH |
| 212 071 | RK | 212 147 | FG | 212 211 | RK | 212 276 | TK | 212 331 | NWH |
| 212 072 | FD | 212 148 | FG | 212 212 | RK | 212 278 | AL | 212 342 | SSH |
| 212 073 | FD | 212 149 | (Z) | 212 213 | RK | 212 279 | AL | 212 343 | SSH |
| 212 074 | RK | 212 150 | (Z) | 212 214 | RK | 212 280 | HO | 212 344 | SSH |
| 212 075 | KM | 212 151 | AL | 212 215 | RK | 212 281 | HO | 212 345 | SSH |
| 212 076 | SSH | 212 152 | MKP | 212 216 | RK | 212 282 | SSH | 212 346 | FD |
| 212 077 | RF | 212 153 | FG | 212 217 | RK | 212 283 | KK | 212 347 | SSH |
| 212 078 | RF | 212 154 | FG | 212 218 | FG | 212 284 | NHO | 212 348 | TK |
| 212 079 | RF | 212 157 | FG | 212 219 | FG | 212 285 | NHO | 212 349 | TK |
| 212 080 | NWH | 212 158 | HBS | 212 221 | MMF | 212 286 | NWH | 212 350 | NWH |
| 212 081 | NWH | 212 159 | HBS | 212 222 | TK | 212 287 | FG | 212 351 | NWH |
| 212 083 | MMF | 212 160 | MKP | 212 223 | RK | 212 288 | KK | 212 354 | FD |
| 212 084 | TK | 212 161 | MKP | 212 224 | TK | 212 289 | FG | 212 355 | FD |
| 212 085 | HG | 212 162 | AL | 212 225 | TK | 212 290 | NHO | 212 356 | FD |
| 212 086 | TK | 212 163 | FG | 212 226 | TK | 212 291 | NHO | 212 357 | FD |
| 212 087 | NWH | 212 165 | FG | 212 227 | TK | 212 292 | FG | 212 358 | FD |
| 212 088 | HBS | 212 166 | RK | 212 228 | SSH | 212 293 | EHG | 212 359 | FD |
| 212 089 | NWH | 212 167 | RK | 212 229 | SSH | 212 294 | NWH | 212 360 | FD |
| 212 091 | RK | 212 168 | (Z) | 212 230 | SSH | 212 295 | NWH | 212 361 | FD |
| 212 093 | MMF | 212 169 | MMF | 212 231 | SSH | 212 296 | AL | 212 364 | FD |
| 212 094 | MKP | 212 170 | MMF | 212 232 | NWH | 212 297 | KK | 212 367 | FD |
| 212 095 | FD | 212 171 | RK | 212 233 | NWH | 212 298 | FG | 212 369 | FD |
| 212 096 | KK | 212 172 | RK | 212 234 | KM | 212 299 | EHG | 212 370 | FD |
| 212 097 | KK | 212 173 | FG | 212 237 | (Z) | 212 300 | NWH | 212 371 | FG |
| 212 098 | (Z) | 212 174 | RK | 212 238 | (Z) | 212 301 | NWH | 212 372 | FG |
| 212 099 | FD | 212 175 | RK | 212 239 | KM | 212 302 | AL | 212 373 | FG |
| 212 100 | MMF | 212 176 | RK | 212 240 | TK | 212 303 | NWH | 212 374 | FG |
| 212 101 | NWH | 212 177 | MKP | 212 241 | NWH | 212 304 | FG | 212 375 | RF |
| 212 102 | NWH | 212 178 | MKP | 212 242 | TK | 212 305 | EHG | 212 376 | SSH |
| 212 103 | NWH | 212 179 | MKP | 212 247 | TK | 212 306 | EHG | 212 377 | SSH |
| 212 104 | FG | 212 180 | MKP | 212 248 | NWH | 212 307 | EHG | 212 380 | NWH |
| 212 105 | FG | 212 181 | MKP | 212 249 | FG | 212 308 | EHG | 212 381 | NWH |
| 212 106 | HBS | 212 182 | TK | | | | | | |

## CLASS 213 B-B

**Built:** 1965-66 by MaK.
**Engine:** Mercedes 12 V 652 TZ of 993 kW at 1500 rpm.
**Transmission:** Hydraulic. Voith L620brs.
**Weight:** 63.00 tonnes. **Maximum Speed:** 100 km/h.
**Length:** 12.300 m. **Train Supply:** Steam.

◀ 201 273 plus one coach form a Schleiz service at Saalburg on 30.04.95
(P. Wormald)

▶ 211 029 heads the 1426 Plattling - Bayerisch Eisenstein at Deggendorf on 03.03.92
(M. Dunn)

◀ 155 018 waits time at Köln Hbf with a Kaldenkirchen - Köln Deutz working on 27.03.95
(P. Wormald)

▶ 171 012 at Huttenrode with the 1148 Blankenburg - Königshutte on 28.09.94
(M. Dunn)

| | | | | |
|---|---|---|---|---|
| 213 332 KK | 213 334 KK | 213 336 UM | 213 338 UM | 213 340 KK |
| 213 333 UM | 213 335 UM | 213 337 UM | 213 339 UM | 213 341 UM |

## CLASS 215          B-B

**Built:** 1968-71 by MaK/Henschel/Krauss-Maffei/Krupp.
**Engine:** Mercedes 16 V 652 TB of 1400 kw (* MAN 12 V 956 TB 10 of 1840 kW) at 1500 rpm.
**Transmission:** Hydraulic. Voith L820brs.
**Weight:** 77.50-79.00 tonnes.      **Maximum Speed:** 140 km/h.
**Length:** 16.400 m.      **Train Supply:** (215 030-032) Electric; (Others) Steam.

| | | | | | | | | | |
|---|---|---|---|---|---|---|---|---|---|
| 215 001* | TU | 215 028 | KK | 215 057 | TU | 215 091* | TU | 215 122 | KK |
| 215 002* | TU | 215 029 | KK | 215 058 | TU | 215 092* | TU | 215 123 | KK |
| 215 003* | TU | 215 030 | TU | 215 059 | TU | 215 093* | TU | 215 124 | KK |
| 215 004* | TU | 215 031 | TU | 215 060 | TU | 215 094 | KK | 215 125 | KK |
| 215 005* | TU | 215 032 | TU | 215 061 | TU | 215 095 | KK | 215 126 | KK |
| 215 006* | TU | 215 033 | STR | 215 062 | TU | 215 096 | KK | 215 127 | KK |
| 215 007* | TU | 215 034 | STR | 215 063 | TU | 215 097 | KK | 215 128 | KK |
| 215 008* | TU | 215 035 | STR | 215 064 | TU | 215 098 | STR | 215 129 | STR |
| 215 009* | TU | 215 036 | STR | 215 065 | TU | 215 099 | STR | 215 130 | KK |
| 215 010* | TU | 215 037 | STR | 215 066 | TU | 215 100 | TU | 215 131 | KK |
| 215 011 | KK | 215 038 | STR | 215 067 | TU | 215 101 | TU | 215 132 | TU |
| 215 012 | KK | 215 039 | STR | 215 068 | TU | 215 105 | TU | 215 133 | STR |
| 215 013 | KK | 215 040 | STR | 215 069 | TU | 215 106 | TU | 215 134 | STR |
| 215 014 | KK | 215 042 | STR | 215 070 | TU | 215 107 | TU | 215 135 | STR |
| 215 015 | KK | 215 043 | STR | 215 071* | TU | 215 109 | TU | 215 136 | STR |
| 215 016 | KK | 215 045 | STR | 215 072* | TU | 215 110 | KK | 215 137 | TU |
| 215 017 | KK | 215 046 | STR | 215 073* | EOB | 215 111 | KK | 215 138 | TU |
| 215 018 | KK | 215 047 | STR | 215 074* | TU | 215 113 | KK | 215 139 | TU |
| 215 019 | KK | 215 048 | STR | 215 075* | EOB | 215 114 | KK | 215 140 | TU |
| 215 020 | KK | 215 049 | KK | 215 076* | EOB | 215 115 | KK | 215 141 | TU |
| 215 021 | KK | 215 050 | TU | 215 077* | EOB | 215 116 | KK | 215 143 | TU |
| 215 022 | EOB | 215 051 | TU | 215 078* | EOB | 215 117 | KK | 215 144 | TU |
| 215 023 | KK | 215 052 | TU | 215 079* | EOB | 215 118 | KK | 215 145 | TU |
| 215 024 | KK | 215 053 | TU | 215 081* | EOB | 215 119 | KK | 215 147 | TU |
| 215 025 | KK | 215 054 | TU | 215 082* | TU | 215 120 | KK | 215 148 | TU |
| 215 026 | KK | 215 055 | TU | 215 084* | TU | 215 121 | KK | 215 150 | TU |
| 215 027 | KK | 215 056 | TU | 215 086* | TU | | | | |

## CLASS 216          B-B

**Built:** 1964-69 by MaK/Henschel/Krauss-Maffei/Krupp/Deutz.
**Engine:** Mercedes 16 V 652 TB of 1400 kW at 1500 rpm.
**Transmission:** Hydraulic. Voith L821rs.
**Weight:** 76.70 tonnes.      **Maximum Speed:** 120 km/h.
**Length:** 16.000 m.      **Train Supply:** Steam.

| | | | | | | | | | |
|---|---|---|---|---|---|---|---|---|---|
| 216 012 | FK | 216 035 | EOB | 216 060 | FG | 216 083 | EOB | 216 105 | EOB |
| 216 013 | (Z) | 216 039 | FG | 216 061 | HO | 216 084 | EOB | 216 106 | EOB |
| 216 014 | EOB | 216 042 | (Z) | 216 062 | HO | 216 088 | HBS | 216 107 | EOB |
| 216 015 | EOB | 216 044 | FG | 216 065 | HO | 216 091 | EOB | 216 108 | EOB |
| 216 016 | (Z) | 216 045 | EOB | 216 066 | HO | 216 092 | EOB | 216 109 | EOB |
| 216 017 | FG | 216 046 | EOB | 216 067 | HO | 216 093 | EOB | 216 110 | EOB |
| 216 018 | EOB | 216 048 | EOB | 216 068 | HBS | 216 094 | EOB | 216 111 | EOB |
| 216 019 | EOB | 216 049 | HO | 216 069 | HBS | 216 095 | FG | 216 112 | EOB |
| 216 020 | FG | 216 050 | FG | 216 072 | HBS | 216 096 | FG | 216 113 | EOB |
| 216 021 | EOB | 216 052 | HO | 216 073 | HBS | 216 097 | FG | 216 114 | EOB |
| 216 022 | EOB | 216 053 | HO | 216 076 | HO | 216 098 | FK | 216 115 | EOB |
| 216 024 | EOB | 216 054 | HO | 216 077 | HOLD | 216 099 | FK | 216 116 | FG |
| 216 025 | EOB | 216 055 | HO | 216 078 | HOLD | 216 100 | FK | 216 117 | FG |
| 216 026 | EOB | 216 056 | HO | 216 079 | HOLD | 216 101 | FK | 216 118 | FG |
| 216 027 | EOB | 216 057 | HO | 216 080 | HOLD | 216 102 | FK | 216 119 | EOB |
| 216 028 | EOB | 216 058 | HO | 216 081 | HOLD | 216 103 | EOB | 216 120 | HOLD |
| 216 034 | EOB | 216 059 | HO | 216 082 | HOLD | 216 104 | EOB | 216 121 | HBS |

# GERMANY

| | | | | | | | | | |
|---|---|---|---|---|---|---|---|---|---|
| 216 122 | HBS | 216 143 | FG | 216 161 | EOB | 216 180 | HBS | 216 201 | HO |
| 216 123 | HOLD | 216 144 | FG | 216 162 | EOB | 216 181 | HBS | 216 203 | FG |
| 216 125 | FG | 216 145 | FG | 216 163 | EOB | 216 182 | HBS | 216 204 | FG |
| 216 126 | FG | 216 146 | EOB | 216 164 | EOB | 216 183 | HBS | 216 205 | FG |
| 216 127 | FG | 216 147 | EOB | 216 165 | FG | 216 184 | HBS | 216 207 | FG |
| 216 128 | FG | 216 148 | EOB | 216 167 | HBS | 216 185 | FK | 216 208 | FG |
| 216 129 | FG | 216 149 | EOB | 216 168 | HBS | 216 186 | FK | 216 209 | FG |
| 216 130 | FK | 216 150 | EOB | 216 169 | HBS | 216 187 | FK | 216 210 | FG |
| 216 131 | FK | 216 151 | EOB | 216 170 | HBS | 216 188 | FK | 216 211 | FG |
| 216 132 | EOB | 216 152 | EOB | 216 171 | HBS | 216 189 | FK | 216 212 | FG |
| 216 133 | EOB | 216 153 | EOB | 216 172 | HBS | 216 190 | FK | 216 213 | FG |
| 216 134 | FK | 216 154 | EOB | 216 173 | HBS | 216 191 | FG | 216 214 | FG |
| 216 135 | FG | 216 155 | EOB | 216 174 | HBS | 216 192 | FG | 216 217 | FG |
| 216 136 | FG | 216 156 | EOB | 216 175 | HBS | 216 193 | FG | 216 219 | FG |
| 216 137 | FK | 216 157 | EOB | 216 176 | HOLD | 216 194 | FG | 216 221 | FG |
| 216 138 | FK | 216 158 | HBS | 216 177 | HBS | 216 195 | FG | 216 222 | FK |
| 216 139 | FG | 216 159 | HBS | 216 178 | HOLD | 216 196 | FG | 216 223 | FK |
| 216 140 | FK | 216 160 | EOB | 216 179 | HBS | 216 197 | FG | 216 224 | FK |
| 216 142 | FK | | | | | | | | |

## CLASS 217                         B-B

**Built:** 1966-68 by Krupp.
**Engine:** Mercedes 16 V 652 TB of 1400 kW at 1500 rpm.
**Transmission:** Hydraulic. Voith L820brs or Mekydro K254B.
**Weight:** 80.10 tonnes.            **Maximum Speed:** 120 km/h.
**Length:** 16.400 m.            **Train Supply:** Electric.

| | | | | | | | | | |
|---|---|---|---|---|---|---|---|---|---|
| 217 003 | NRH | 217 013 | NRH | 217 016 | NRH | 217 019 | NRH | 217 021 | NRH |
| 217 011 | NRH | 217 014 | NRH | 217 017 | NRH | 217 020 | NRH | 217 022 | NRH |
| 217 012 | NRH | 217 015 | NRH | 217 018 | NRH | | | | | |

## CLASS 218                         B-B

**Built:** 1968-79 by Krupp/Krauss-Maffei/Henschel/MaK.
**Engine:** MAN 12 V 956 TB 10 of 1840 kW (* MAN 12 V 956 TB 11 of 2060 kW; † Pielstick 16 PA 4V 200 of 2060 kW) all at 1500 rpm.
**Transmission:** Hydraulic. Voith L820brs or Mekydro K252SUBB.
**Weight:** 78.70 tonnes.            **Maximum Speed:** 140 km/h.
**Length:** 16.400 m.            **Train Supply:** Electric.

| | | | | | | | | | |
|---|---|---|---|---|---|---|---|---|---|
| 218 001 | NRH | 218 112 | AFL | 218 135 | EHG | 218 158 | AL | 218 181 | AL |
| 218 002 | NRH | 218 113 | AFL | 218 136 | EHG | 218 159 | AL | 218 182 | AL |
| 218 003 | NRH | 218 114 | AFL | 218 137 | EHG | 218 160 | AL | 218 183 | AL |
| 218 004 | NRH | 218 115 | AFL | 218 138 | EHG | 218 161 | AL | 218 184 | AL |
| 218 005 | NRH | 218 116 | AFL | 218 139 | EHG | 218 162 | AL | 218 185 | AL |
| 218 006 | NRH | 218 117 | AFL | 218 140 | EHG | 218 163 | AL | 218 186 | AL |
| 218 007 | NRH | 218 118 | AFL | 218 141 | EHG | 218 164 | AL | 218 187 | AL |
| 218 008 | NRH | 218 119 | AFL | 218 142 | EHG | 218 165 | AL | 218 188 | AL |
| 218 009 | NRH | 218 120 | AFL | 218 143 | EHG | 218 166 | AL | 218 189 | AL |
| 218 010 | NRH | 218 121 | AFL | 218 144 | EHG | 218 167 | AFL | 218 190 | AL |
| 218 011 | NRH | 218 122 | AFL | 218 145 | EHG | 218 168 | AFL | 218 191 | AL |
| 218 012 | NRH | 218 123 | AFL | 218 146 | EHG | 218 169 | AFL | 218 192 | AL |
| 218 101 | AFL | 218 124 | AFL | 218 147 | HBS | 218 170 | AFL | 218 193 | AL |
| 218 102 | AFL | 218 125 | AFL | 218 148 | EHG | 218 171 | AFL | 218 194 | AL |
| 218 103 | AFL | 218 126 | AFL | 218 149 | EHG | 218 172 | AFL | 218 195 | AL |
| 218 104 | AFL | 218 127 | AFL | 218 150 | EHG | 218 173 | AFL | 218 196 | AL |
| 218 105 | AFL | 218 128 | EHG | 218 151 | AL | 218 174 | AFL | 218 197† | AL |
| 218 106 | AFL | 218 129 | EHG | 218 152 | AL | 218 175 | AFL | 218 198* | RHL |
| 218 107 | AFL | 218 130 | EHG | 218 153 | AL | 218 176 | AL | 218 199* | RHL |
| 218 108 | AFL | 218 131 | EHG | 218 154 | AL | 218 177 | AL | 218 200* | RK |
| 218 109 | AFL | 218 132 | EHG | 218 155 | AL | 218 178 | AL | 218 201* | RK |
| 218 110 | AFL | 218 133 | EHG | 218 156 | AL | 218 179 | AL | 218 202* | HBS |
| 218 111 | AFL | 218 134 | EHG | 218 157 | AL | 218 180 | AL | 218 203* | HBS |

| No. | Code | No. | Code | No. | Code | No. | Code | No. | Code |
|---|---|---|---|---|---|---|---|---|---|
| 218 204* | HBS | 218 266 | HBS | 218 328† | AL | 218 389* | RHL | 218 450* | MKP |
| 218 205* | HBS | 218 268 | HBS | 218 329† | AL | 218 390* | MMF | 218 451* | MKP |
| 218 206* | HBS | 218 269 | HBS | 218 330† | AL | 218 391* | MMF | 218 452* | NRH |
| 218 207* | NRH | 218 270 | HBS | 218 331† | AL | 218 392* | MMF | 218 453* | NRH |
| 218 208* | HBS | 218 271 | HBS | 218 332† | AL | 218 393* | MMF | 218 454* | NRH |
| 218 209* | NRH | 218 272 | HBS | 218 333† | AL | 218 394* | RK | 218 455* | MKP |
| 218 210* | NRH | 218 273 | HBS | 218 334† | AL | 218 395* | RK | 218 456† | AL |
| 218 211* | NRH | 218 274 | HBS | 218 335† | AL | 218 396* | RHL | 218 457† | AL |
| 218 212* | NRH | 218 275 | HBS | 218 336† | AL | 218 397* | RHL | 218 458† | AL |
| 218 213* | NRH | 218 276 | HBS | 218 337† | AL | 218 398* | MKP | 218 459† | AL |
| 218 214* | NRH | 218 277 | HBS | 218 338† | AL | 218 399* | MKP | 218 460† | AL |
| 218 215* | NRH | 218 278 | HBS | 218 339† | AL | 218 400* | MKP | 218 461† | AL |
| 218 216* | NRH | 218 279 | AL | 218 340* | MMF | 218 401* | MKP | 218 462† | AL |
| 218 217* | NRH | 218 280 | AL | 218 341* | MMF | 218 402* | MKP | 218 463* | MKP |
| 218 218* | NRH | 218 281 | AL | 218 342* | MMF | 218 403* | NRH | 218 464* | MKP |
| 218 219* | NRH | 218 282 | AL | 218 343* | MMF | 218 404* | NRH | 218 465* | MKP |
| 218 220* | NRH | 218 283 | AL | 218 344* | MMF | 218 405* | NRH | 218 466* | MKP |
| 218 221* | NRH | 218 284 | AL | 218 345* | MMF | 218 406* | NRH | 218 467* | MKP |
| 218 222* | MKP | 218 285 | AL | 218 346* | MMF | 218 407* | NRH | 218 468* | MKP |
| 218 223* | MKP | 218 286 | AL | 218 347* | MMF | 218 408* | NRH | 218 469* | MKP |
| 218 224* | MKP | 218 287 | AL | 218 348* | MMF | 218 409* | NRH | 218 470* | MKP |
| 218 225* | MKP | 218 288 | HBS | 218 349* | MMF | 218 410* | NRH | 218 471* | MKP |
| 218 226* | MKP | 218 289* | RHL | 218 350* | MMF | 218 411* | NRH | 218 472* | MKP |
| 218 227* | MKP | 218 290* | RHL | 218 351* | MMF | 218 412* | NRH | 218 473* | MKP |
| 218 228* | MKP | 218 291* | RHL | 218 352* | MMF | 218 413* | NRH | 218 474* | MKP |
| 218 229* | MKP | 218 292* | RHL | 218 353* | MMF | 218 414* | NRH | 218 475* | MKP |
| 218 230* | MKP | 218 293* | RHL | 218 354* | MMF | 218 415* | NRH | 218 476* | RK |
| 218 231* | MKP | 218 294* | RK | 218 355* | MMF | 218 416* | NRH | 218 477* | RK |
| 218 232* | MKP | 218 295* | RK | 218 356* | MMF | 218 417* | NRH | 218 478* | RK |
| 218 233* | MKP | 218 296* | RK | 218 357* | MMF | 218 418* | NRH | 218 479* | RHL |
| 218 234* | MKP | 218 297* | RK | 218 358* | MMF | 218 419* | NRH | 218 480* | RHL |
| 218 235* | MKP | 218 298* | RK | 218 359† | MMF | 218 420* | NRH | 218 481* | RHL |
| 218 236* | MKP | 218 299* | RK | 218 360* | MMF | 218 421* | NRH | 218 482* | RHL |
| 218 237* | MKP | 218 300* | RK | 218 361* | SKL | 218 422* | NRH | 218 483* | RHL |
| 218 238* | MKP | 218 301* | RK | 218 362* | SKL | 218 423* | NRH | 218 484* | RHL |
| 218 239* | MKP | 218 302* | RK | 218 363* | SKL | 218 424* | MKP | 218 485† | AL |
| 218 240* | MKP | 218 303* | RK | 218 364* | SKL | 218 425* | MKP | 218 486† | AL |
| 218 241* | MKP | 218 304* | RK | 218 365* | SKL | 218 426* | MKP | 218 487† | AL |
| 218 242 | HBS | 218 305* | RHL | 218 366* | SKL | 218 427† | AL | 218 488† | AL |
| 218 244 | AL | 218 306* | RHL | 218 367* | SKL | 218 428† | AL | 218 489† | AL |
| 218 245 | HBS | 218 307* | MMF | 218 368* | SKL | 218 429† | AL | 218 490† | AL |
| 218 246 | HBS | 218 308* | MMF | 218 369* | SKL | 218 430† | AL | 218 491† | AL |
| 218 247 | EHG | 218 309* | MMF | 218 370* | SKL | 218 431† | AL | 218 492† | AL |
| 218 248 | HBS | 218 310* | MMF | 218 371* | SKL | 218 432† | AL | 218 493† | AL |
| 218 249 | EHG | 218 311* | MMF | 218 372* | SKL | 218 433† | AL | 218 494† | AL |
| 218 250 | HBS | 218 312* | MMF | 218 373* | SKL | 218 434† | AL | 218 495† | AL |
| 218 251 | HBS | 218 313* | MMF | 218 374* | SKL | 218 435* | MKP | 218 496† | AL |
| 218 252 | HBS | 218 314* | MMF | 218 375* | SKL | 218 436* | MKP | 218 497† | AL |
| 218 253 | HBS | 218 315* | MMF | 218 376* | SKL | 218 437* | MKP | 218 498† | AL |
| 218 254 | AL | 218 316* | MMF | 218 377* | SKL | 218 438* | MKP | 218 499† | AL |
| 218 255 | HBS | 218 317* | MMF | 218 378* | SKL | 218 439* | MKP | 218 901 | HBS |
| 218 256 | AL | 218 318* | MMF | 218 379* | SKL | 218 440* | MKP | 218 902 | HBS |
| 218 257 | AL | 218 319* | MMF | 218 380* | SKL | 218 441* | MKP | 218 903 | HBS |
| 218 258 | AL | 218 320* | MMF | 218 381* | SKL | 218 442* | MKP | 218 904 | HBS |
| 218 259 | AL | 218 321* | MMF | 218 382* | MMF | 218 443* | MKP | 218 905 | HBS |
| 218 260 | AL | 218 322* | MMF | 218 383* | MMF | 218 444* | MKP | 218 906 | HBS |
| 218 261 | AL | 218 323† | AL | 218 384* | SKL | 218 445* | MKP | 218 907 | HBS |
| 218 262 | HBS | 218 324† | AL | 218 385* | SKL | 218 446* | MKP | 218 908 | HBS |
| 218 263 | HBS | 218 325† | AL | 218 386* | SKL | 218 447* | MKP | | |
| 218 264 | HBS | 218 326† | AL | 218 387* | SKL | 218 448* | MKP | | |
| 218 265 | HBS | 218 327† | AL | 218 388* | RHL | 218 449* | MKP | | |

# CLASS 219      C-C

**Built:** 1976-85 by 23rd August.
**Engines:** Two Mercedes 820SR of 990 kW at 1500 rpm.
**Transmission:** Hydraulic. Strömungsmachinen GS 30/5.5.
**Weight:** 96.00 tonnes.      **Maximum Speed:** 120 km/h.
**Length:** 19.500 m.      **Train Supply:** Electric.
**Note:** Previously DR Class 119.

| | | | | | | | | | |
|---|---|---|---|---|---|---|---|---|---|
| 219 003 | DC | 219 043 | LS | 219 078 | DG | 219 117 | US | 219 157 | US |
| 219 004 | LS | 219 044 | (Z) | 219 079 | US | 219 119 | US | 219 158 | LS |
| 219 005 | WS | 219 045 | DG | 219 080 | DG | 219 121 | DC | 219 159 | US |
| 219 006 | US | 219 046 | DG | 219 081 | WS | 219 122 | LS | 219 160 | DG |
| 219 007 | DG | 219 047 | LS | 219 082 | LS | 219 123 | DG | 219 161 | US |
| 219 009 | WS | 219 048 | DG | 219 083 | WS | 219 124 | US | 219 162 | LS |
| 219 010 | DG | 219 049 | LS | 219 084 | US | 219 125 | WS | 219 163 | LS |
| 219 012 | DG | 219 050 | LS | 219 085 | US | 219 127 | DC | 219 164 | DG |
| 219 013 | WS | 219 051 | DC | 219 086 | US | 219 129 | WS | 219 165 | DC |
| 219 014 | LS | 219 052 | DG | 219 087 | US | 219 130 | DG | 219 166 | DG |
| 219 015 | DG | 219 053 | US | 219 089 | WS | 219 131 | DC | 219 167 | WS |
| 219 016 | DG | 219 054 | DG | 219 090 | US | 219 132 | US | 219 168 | DC |
| 219 017 | WSR | 219 055 | DC | 219 091 | DC | 219 133 | US | 219 169 | (Z) |
| 219 018 | WS | 219 056 | US | 219 092 | DG | 219 134 | DG | 219 175 | US |
| 219 020 | US | 219 057 | WS | 219 093 | DG | 219 135 | US | 219 176 | DG |
| 219 021 | US | 219 059 | US | 219 095 | WS | 219 136 | DC | 219 177 | DG |
| 219 022 | DG | 219 060 | LS | 219 096 | WS | 219 137 | US | 219 178 | US |
| 219 023 | DG | 219 061 | US | 219 097 | WS | 219 138 | US | 219 179 | DG |
| 219 024 | DG | 219 062 | LS | 219 098 | US | 219 139 | US | 219 180 | US |
| 219 025 | DC | 219 063 | US | 219 099 | WS | 219 141 | WS | 219 182 | (Z) |
| 219 026 | DG | 219 064 | DG | 219 101 | US | 219 142 | DG | 219 183 | LS |
| 219 027 | US | 219 065 | DG | 219 103 | WS | 219 143 | WS | 219 185 | LS |
| 219 029 | US | 219 066 | LS | 219 104 | US | 219 145 | DG | 219 187 | DG |
| 219 030 | US | 219 067 | DC | 219 105 | DC | 219 146 | DG | 219 189 | LS |
| 219 032 | LS | 219 068 | (Z) | 219 107 | DG | 219 148 | WS | 219 190 | US |
| 219 034 | (Z) | 219 069 | WS | 219 108 | US | 219 149 | DC | 219 191 | LS |
| 219 035 | DG | 219 070 | DC | 219 109 | WS | 219 150 | DG | 219 192 | DG |
| 219 036 | LS | 219 071 | DC | 219 110 | DG | 219 151 | US | 219 194 | US |
| 219 037 | DC | 219 072 | US | 219 111 | DC | 219 152 | US | 219 195 | US |
| 219 038 | DG | 219 073 | (Z) | 219 112 | US | 219 153 | LS | 219 196 | (Z) |
| 219 039 | US | 219 074 | DG | 219 114 | LS | 219 154 | DC | 219 197 | WS |
| 219 040 | LS | 219 075 | DG | 219 115 | US | 219 155 | WS | 219 198 | LS |
| 219 041 | DG | 219 076 | US | 219 116 | WS | 219 156 | DC | 219 200 | US |
| 219 042 | (Z) | 219 077 | DC | | | | | | |

# CLASS 220      Co-Co

**Built:** 1966-75 by Voroshilovgrad.
**Engine:** Kolomna 14 D 40 of 1470 kW at 750 rpm.
**Transmission:** Electric. Charkov ED107 traction motors.
**Weight:** 116.00 tonnes.      **Maximum Speed:** 100 km/h.
**Length:** 17.550 m.      **Train Supply:** Not equipped.
**Note:** Previously DR Class 120.

| | | | | | | | | | |
|---|---|---|---|---|---|---|---|---|---|
| 220 086 | (Z) | 220 216 | (Z) | 220 273 | (Z) | 220 332 | (Z) | 220 336 | (Z) |
| 220 124 | (Z) | 220 241 | (Z) | 220 296 | (Z) | 220 334 | (Z) | 220 339 | (Z) |
| 220 185 | (Z) | 220 272 | (Z) | | | | | | |

# CLASS 228.2      C-C

**Built:** 1966-70 by LKM.
**Engines:** Two Johannisthal 12 KVD 21 AL4 (* 12 KVD 18/21 A2) of 883 (* 736) kW at 1500 rpm.
**Transmission:** Hydraulic. Strömungsmachinen GSR 30/5.7 AQ or L306rb.
**Weight:** 90.00 tonnes.      **Maximum Speed:** 120 km/h.
**Length:** 19.460 m.      **Train Supply:** Steam.
**Note:** Previously DR Class 118.

| | | | | |
|---|---|---|---|---|
| 228 372 LMR | 228 680 (Z) | 228 706 (Z) | 228 756 (Z) | 228 788 UE |
| 228 608 (Z) | 228 683 (Z) | 228 708 (Z) | 228 761 (Z) | 228 791 UE |
| 228 614 (Z) | 228 685 (Z) | 228 710 (Z) | 228 764 (Z) | 228 792 (Z) |
| 228 616 (Z) | 228 686 (Z) | 228 715 (Z) | 228 766 UE | 228 794 (Z) |
| 228 621 (Z) | 228 688 (Z) | 228 724 (Z) | 228 767 UE | 228 795 UE |
| 228 623 (Z) | 228 692 LMR | 228 725 (Z) | 228 769 UE | 228 798 UE |
| 228 631 (Z) | 228 693 LMR | 228 736 (Z) | 228 771 (Z) | 228 800 (Z) |
| 228 640 LMR | 228 695 (Z) | 228 739 (Z) | 228 773 (Z) | 228 801 UE |
| 228 646 UE | 228 696 (Z) | 228 740 (Z) | 228 777 (Z) | 228 802 (Z) |
| 228 662 (Z) | 228 697 (Z) | 228 746 UE | 228 780 (Z) | 228 803 (Z) |
| 228 671 (Z) | 228 700 (Z) | 228 748 UE | 228 782 (Z) | 228 804 LMR |
| 228 674 (Z) | 228 703 (Z) | 228 749 UE | 228 784 UE | 228 805 (Z) |
| 228 675 (Z) | 228 704 (Z) | 228 751 UE | 228 786 UE | 228 806 UE |

# CLASS 229 — C-C

**Built:** 1976-85 by 23rd August. Rebuilt 1992-93 by Krupp.
**Engines:** Two MTU of 1380 kW.
**Transmission:** Hydraulic.
**Weight:**
**Length:** 19.500 m.
**Note:** Previously DR Class 119.

**Maximum Speed:** 140 km/h.
**Train Supply:** Electric.

| | | | | |
|---|---|---|---|---|
| 229 100 UE | 229 118 UE | 229 144 UE | 229 173 UE | 229 186 UE |
| 229 102 UE | 229 120 UE | 229 147 UE | 229 174 UE | 229 188 UE |
| 229 106 UE | 229 126 UE | 229 170 UE | 229 181 UE | 229 193 UE |
| 229 113 UE | 229 128 UE | 229 171 UE | 229 184 UE | 229 199 UE |

# CLASS 232 — Co-Co

**Built:** 1973-82 by Voroshilovgrad.
**Engine:** Kolomna 5 D 49 of 2200 kW at 1000 rpm.
**Transmission:** Electric. Charkov ED118 traction motors.
**Weight:** 122.00 tonnes.
**Length:** 20.820 m.
**Note:** Previously DR Class 132.

**Maximum Speed:** 120 km/h.
**Train Supply:** Electric.

| | | | | | |
|---|---|---|---|---|---|
| 232 002 BSE | 232 032 LM | 232 061 DG | 232 093 LL1 | 232 127 UE | |
| 232 003 UE | 232 033 LHB | 232 062 LM | 232 094 WP | 232 128 LM | |
| 232 004 BSE | 232 034 UE | 232 064 LM | 232 096 DRC | 232 129 WS | |
| 232 005 LM | 232 035 UE | 232 065 BCS | 232 097 LM | 232 130 US | |
| 232 006 (Z) | 232 036 BPKR | 232 066 BCS | 232 098 LL1 | 232 131 WS | |
| 232 008 WS | 232 037 BSE | 232 068 (Z) | 232 099 DC | 232 132 WS | |
| 232 009 UE | 232 038 US | 232 070 UM | 232 100 UEI | 232 134 BPKR | |
| 232 010 BCS | 232 039 LM | 232 071 DG | 232 101 WP | 232 135 UE | |
| 232 011 BCS | 232 040 UE | 232 073 BSE | 232 104 DG | 232 136 UE | |
| 232 012 UM | 232 041 DG | 232 074 BHW | 232 105 LM | 232 137 WS | |
| 232 014 BSE | 232 042 BHW | 232 076 UEI | 232 106 UEI | 232 138 DG | |
| 232 017 LH1 | 232 043 WS | 232 077 DZW | 232 107 (Z) | 232 140 DRC | |
| 232 018 DRC | 232 044 UEI | 232 078 UN | 232 108 UEI | 232 141 US | |
| 232 019 WP | 232 045 WS | 232 079 DRC | 232 109 UM | 232 142 DG | |
| 232 020 UE | 232 046 UEI | 232 080 DG | 232 110 (Z) | 232 143 US | |
| 232 021 LM | 232 047 DC | 232 081 DG | 232 111 LM | 232 145 UEI | |
| 232 022 BCS | 232 048 UM | 232 083 DRC | 232 112 UN | 232 146 LM | |
| 232 023 (Z) | 232 049 BSE | 232 084 LMR | 232 113 US | 232 147 BSE | |
| 232 024 UM | 232 050 BSE | 232 085 DRC | 232 114 UN | 232 148 UEI | |
| 232 025 UM | 232 051 LM | 232 086 UEI | 232 117 LL1 | 232 149 LM | |
| 232 026 DRC | 232 052 UEI | 232 087 UM | 232 118 WS | 232 151 WS | |
| 232 027 WS | 232 053 DRC | 232 088 BSE | 232 120 US | 232 152 DG | |
| 232 028 LM | 232 054 BHW | 232 089 UM | 232 121 UN | 232 153 WP | |
| 232 029 BCS | 232 055 BSE | 232 090 US | 232 122 LL1 | 232 154 LM | |
| 232 030 UE | 232 059 LH1 | 232 091 DRC | 232 123 BSE | 232 155 WS | |
| 232 031 BCS | 232 060 UEI | 232 092 DRC | 232 125 WS | 232 156 WS | |

# GERMANY

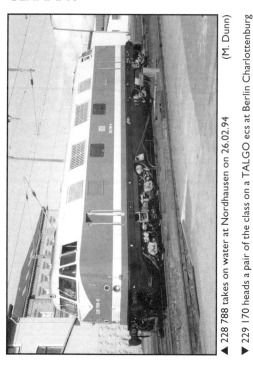

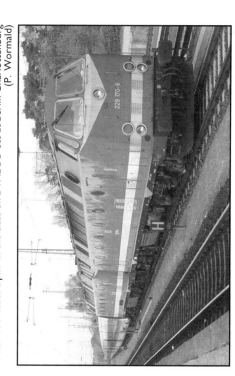

▲ 228 788 takes on water at Nordhausen on 26.02.94 (M. Dunn)

▼ 229 170 heads a pair of the class on a TALGO ecs at Berlin Charlottenburg (P. Wormald)

▲ 216 210 awaits departure from Gießen with the 1155 to Koblenz on 05.05.94 (M. Dunn)

▼ 218 205 and another unidentified 218 at Reichenbach depot on 09.02.95 (P. Wormald)

| | | | | | | | | | |
|---|---|---|---|---|---|---|---|---|---|
| 232 157 | DG | 232 239 | LHB | 232 325 | (Z) | 232 406 | UE | 232 483 | DG |
| 232 158 | DG | 232 240 | DRC | 232 326 | EOB | 232 407 | BCS | 232 484 | LM |
| 232 162 | DG | 232 241 | DG | 232 328 | WP | 232 408 | UM | 232 485 | DRC |
| 232 164 | BCS | 232 245 | LHB | 232 329 | DG | 232 409 | BCS | 232 486 | BSE |
| 232 165 | DRC | 232 246 | DG | 232 330 | LHB | 232 410 | LM | 232 487 | UE |
| 232 167 | DRC | 232 248 | LL1 | 232 331 | UEI | 232 411 | LM | 232 488 | UEI |
| 232 168 | UEI | 232 249 | LM | 232 332 | LHB | 232 412 | BCS | 232 489 | UE |
| 232 172 | LM | 232 250 | LH1 | 232 333 | LL1 | 232 413 | LH1 | 232 491 | UN |
| 232 173 | LL1 | 232 252 | BCS | 232 334 | WS | 232 414 | LL1 | 232 492 | UN |
| 232 174 | LM | 232 253 | WS | 232 337 | LHB | 232 415 | UW | 232 493 | BCS |
| 232 176 | LMR | 232 254 | LH1 | 232 338 | UE | 232 416 | WS | 232 494 | LHB |
| 232 177 | (Z) | 232 255 | LL1 | 232 339 | WP | 232 418 | BPKR | 232 495 | BCS |
| 232 178 | LH1 | 232 256 | UEI | 232 342 | BCS | 232 419 | WNT | 232 496 | BSE |
| 232 179 | UE | 232 258 | LHB | 232 345 | UM | 232 420 | LHB | 232 497 | DG |
| 232 181 | UE | 232 259 | LH1 | 232 347 | LHB | 232 421 | UE | 232 498 | LM |
| 232 182 | DRC | 232 260 | UE | 232 349 | LHB | 232 424 | LM | 232 499 | WS |
| 232 184 | WS | 232 261 | WP | 232 350 | BPKR | 232 425 | US | 232 500 | DC |
| 232 186 | BCS | 232 262 | LL1 | 232 352 | WS | 232 426 | UEI | 232 501 | WS |
| 232 187 | LHB | 232 263 | US | 232 353 | UE | 232 427 | WP | 232 502 | LM |
| 232 188 | DG | 232 264 | US | 232 354 | DRC | 232 428 | BSE | 232 503 | BCS |
| 232 189 | LM | 232 265 | LHB | 232 355 | (Z) | 232 429 | WNT | 232 505 | UM |
| 232 190 | BSE | 232 268 | LH1 | 232 356 | LH1 | 232 430 | DG | 232 506 | UW |
| 232 191 | LM | 232 269 | LL1 | 232 357 | WS | 232 432 | UEI | 232 508 | WS |
| 232 192 | DG | 232 271 | DRC | 232 358 | LL1 | 232 434 | WP | 232 509 | WP |
| 232 194 | US | 232 272 | BSE | 232 359 | EOB | 232 435 | LL1 | 232 510 | DRC |
| 232 195 | WS | 232 273 | LL1 | 232 361 | UM | 232 436 | WP | 232 511 | EOB |
| 232 196 | LHB | 232 274 | LHB | 232 362 | UE | 232 437 | UEI | 232 512 | BSE |
| 232 197 | UN | 232 276 | LL1 | 232 363 | BCS | 232 438 | US | 232 513 | WS |
| 232 198 | LM | 232 277 | DG | 232 365 | WS | 232 439 | WP | 232 514 | WS |
| 232 201 | UEI | 232 280 | DRC | 232 367 | EOB | 232 441 | UE | 232 515 | UEI |
| 232 202 | BSE | 232 281 | EOB | 232 368 | LL1 | 232 443 | UE | 232 516 | (Z) |
| 232 203 | LM | 232 282 | BCS | 232 371 | BPKR | 232 445 | WP | 232 517 | WP |
| 232 204 | UEI | 232 283 | UM | 232 372 | WS | 232 446 | DRC | 232 518 | BHW |
| 232 205 | UE | 232 284 | BCS | 232 373 | WS | 232 447 | (Z) | 232 519 | LL1 |
| 232 206 | UE | 232 285 | UE | 232 374 | DC | 232 448 | LL1 | 232 520 | UN |
| 232 207 | UEI | 232 286 | BPKR | 232 375 | UN | 232 449 | DRC | 232 521 | EOB |
| 232 208 | US | 232 287 | WP | 232 376 | US | 232 450 | LL1 | 232 524 | UEI |
| 232 209 | LL1 | 232 288 | EOB | 232 377 | BSE | 232 451 | EOB | 232 525 | EOB |
| 232 211 | LH1 | 232 289 | EOB | 232 378 | BSE | 232 452 | EOB | 232 527 | LM |
| 232 212 | UM | 232 290 | LL1 | 232 379 | UE | 232 453 | LHB | 232 528 | UEI |
| 232 213 | UE | 232 291 | UN | 232 380 | DG | 232 454 | LHB | 232 529 | LL1 |
| 232 215 | UN | 232 293 | WP | 232 381 | UM | 232 455 | LL1 | 232 530 | UE |
| 232 216 | UE | 232 294 | WS | 232 382 | BPKR | 232 456 | DG | 232 531 | LM |
| 232 217 | UE | 232 295 | LM | 232 383 | LL1 | 232 457 | WP | 232 532 | BSE |
| 232 218 | DRC | 232 296 | WS | 232 384 | LL1 | 232 458 | LHB | 232 533 | DG |
| 232 219 | LM | 232 297 | WP | 232 386 | BCS | 232 459 | LM | 232 534 | WS |
| 232 220 | BPKR | 232 298 | BSE | 232 387 | WP | 232 460 | BCS | 232 535 | WS |
| 232 221 | DG | 232 300 | LL1 | 232 388 | LH1 | 232 461 | UE | 232 536 | LHB |
| 232 222 | DG | 232 301 | BCS | 232 390 | LL1 | 232 462 | UE | 232 537 | DRC |
| 232 223 | LM | 232 303 | LM | 232 391 | WS | 232 464 | DRC | 232 539 | BSE |
| 232 225 | LM | 232 305 | BSE | 232 392 | BSE | 232 465 | WS | 232 540 | BSE |
| 232 226 | LHB | 232 306 | EOB | 232 393 | LHB | 232 466 | LHB | 232 541 | WS |
| 232 227 | BCS | 232 308 | WS | 232 394 | WP | 232 469 | EOB | 232 542 | DG |
| 232 228 | BSE | 232 309 | WS | 232 395 | LHB | 232 470 | WS | 232 543 | DRC |
| 232 229 | LH1 | 232 310 | WP | 232 396 | WNT | 232 471 | DG | 232 544 | LM |
| 232 230 | WS | 232 312 | BSE | 232 397 | WS | 232 472 | UEI | 232 545 | BSE |
| 232 231 | LL1 | 232 313 | BHW | 232 398 | LL1 | 232 474 | DG | 232 546 | BHW |
| 232 232 | UE | 232 314 | WS | 232 400 | DRC | 232 476 | DRC | 232 547 | WP |
| 232 233 | WS | 232 315 | US | 232 401 | UEI | 232 477 | WP | 232 549 | BSE |
| 232 234 | DG | 232 317 | LM | 232 402 | UM | 232 478 | EOB | 232 550 | BHW |
| 232 236 | LHB | 232 319 | DG | 232 403 | BCS | 232 480 | DRC | 232 553 | UE |
| 232 237 | DG | 232 321 | LM | 232 404 | BSE | 232 481 | UEI | 232 557 | BHW |
| 232 238 | LHB | 232 322 | EOB | 232 405 | (Z) | 232 482 | US | 232 558 | LL1 |

# GERMANY

| | | | | | | | | | |
|---|---|---|---|---|---|---|---|---|---|
| 232 559 | LM | 232 592 | WS | 232 624 | WS | 232 655 | WP | 232 684 | BCS |
| 232 560 | BPKR | 232 593 | UEI | 232 625 | LM | 232 656 | DG | 232 685 | BCS |
| 232 561 | LM | 232 594 | DC | 232 626 | DRC | 232 658 | UEI | 232 686 | US |
| 232 562 | US | 232 595 | WS | 232 627 | LL1 | 232 659 | LH1 | 232 687 | LL1 |
| 232 563 | DG | 232 596 | BSE | 232 628 | WP | 232 660 | BSE | 232 688 | WNT |
| 232 564 | LM | 232 598 | WS | 232 629 | LL1 | 232 661 | BCS | 232 689 | UE |
| 232 566 | LHB | 232 600 | DRC | 232 631 | BSE | 232 662 | LM | 232 690 | BCS |
| 232 567 | WS | 232 601 | DRC | 232 632 | WP | 232 663 | UE | 232 691 | DRC |
| 232 568 | BPKR | 232 602 | BHW | 232 633 | WP | 232 665 | BCS | 232 692 | BHW |
| 232 569 | LL1 | 232 603 | WS | 232 634 | WS | 232 666 | US | 232 693 | LL1 |
| 232 570 | DRC | 232 604 | UE | 232 635 | LL1 | 232 667 | WSR | 232 694 | LH1 |
| 232 571 | US | 232 605 | US | 232 636 | WS | 232 668 | LH1 | 232 695 | DRC |
| 232 572 | DC | 232 607 | BSE | 232 638 | WP | 232 669 | UM | 232 696 | WS |
| 232 573 | DG | 232 609 | DG | 232 640 | WS | 232 670 | BSE | 232 697 | UN |
| 232 575 | LH1 | 232 610 | BSE | 232 642 | DG | 232 672 | LL1 | 232 698 | LM |
| 232 576 | UEI | 232 611 | UEI | 232 643 | WS | 232 673 | LL1 | 232 699 | WS |
| 232 577 | UEI | 232 612 | US | 232 644 | DG | 232 674 | LL1 | 232 700 | DC |
| 232 579 | LM | 232 613 | BSE | 232 645 | LL1 | 232 675 | LH1 | 232 701 | DRC |
| 232 580 | DG | 232 614 | UEI | 232 646 | UEI | 232 676 | LM | 232 702 | WS |
| 232 581 | BHW | 232 615 | UE | 232 647 | WNT | 232 677 | LH1 | 232 703 | LH1 |
| 232 583 | DG | 232 616 | UN | 232 648 | DRC | 232 678 | DRC | 232 704 | US |
| 232 584 | DRC | 232 617 | DRC | 232 649 | WP | 232 679 | LH1 | 232 705 | BSE |
| 232 586 | BSE | 232 618 | BCS | 232 650 | DG | 232 680 | LM | 232 706 | WS |
| 232 587 | LL1 | 232 620 | WSR | 232 652 | BSE | 232 681 | WNT | 232 707 | WS |
| 232 588 | WS | 232 622 | LH1 | 232 653 | LM | 232 682 | DRC | 232 708 | LM |
| 232 589 | US | 232 623 | US | 232 654 | LM | 232 683 | LH1 | 232 709 | WS |
| 232 590 | UE | | | | | | | | |

## CLASS 234                                                    Co-Co

**Built:** 1973-82 by Voroshilovgrad. Rebuilt 1991 by DR Cottbus Works.
**Engine:** Kolomna 5D49 of 2200 kW at 1000 rpm.
 † Kolomna 12D49 of 2450 kW.
 * Caterpillar 3608 of 2460 kW.
 ‡ MaK 12 M 282 L of 2460 kW.
**Transmission:** Electric. Charkov ED118 traction motors.
**Weight:** 122.00 (*‡ 125.00) tonnes.    **Maximum Speed:** 140 km/h.
**Length:** 20.820 m.    **Train Supply:** Electric.
**Note:** Previously DR Class 232.

| | | | | | | | | | |
|---|---|---|---|---|---|---|---|---|---|
| 234 016 | BWUR | 234 247 | LM | 234 344 | BPKR | 234 504 | DRC | 234 582 | DRC |
| 234 072 | DRC | 234 251 | DRC | 234 346 | BPKR | 234 507 | BPKR | 234 585 | BPKR |
| 234 075 | BPKR | 234 257 | DG | 234 351 | DG | 234 523 | LM | 234 591 | BPKR |
| 234 116 | LM | 234 278 | BPKR | 234 385 | LM | 234 526 | DRC | 234 597 | DG |
| 234 144 | DG | 234 292 | LM | 234 399 | BPKR | 234 538 | DRC | 234 606 | DRC |
| 234 161 | DRC | 234 299 | BPKR | 234 417 | LM | 234 548* | BPKR | 234 608 | DRC |
| 234 166 | DG | 234 304 | DG | 234 423 | DRC | 234 551 | BWUR | 234 630‡ | BPKR |
| 234 170 | DRC | 234 311 | DRC | 234 440† | BPKR | 234 552 | DG | 234 641 | DRC |
| 234 180 | BWUR | 234 320 | BPKR | 234 442 | BPKR | 234 554 | BPKR | 234 651 | BPKR |
| 234 235 | BPKR | 232 323 | DG | 234 467 | BPKR | 234 555‡ | BPKR | 234 657† | BPKR |
| 234 242 | DG | 234 335 | BPKR | 234 468 | DG | 234 565* | BPKR | 234 664 | BPKR |
| 234 244 | BPKR | 234 341 | BPKR | 234 475 | BPKR | 234 578 | BPKR | | |

## CLASS 240                                                    Co-Co

**Built:** 1989 by MaK.
**Engine:** MaK 12 M 282 of 2650 kW at 1000 rpm.
**Transmission:** Electric. ABB.
**Weight:** 122.00 tonnes.    **Maximum Speed:** 160 km/h.
**Length:** 20.960 m.    **Train Supply:** Electric.
**Note:** Test locomotives owned by MaK. Currently off loan, but expected to return to DBAG shortly

| | |
|---|---|
| 240 001 | *Kiel* |
| 240 002 | *Westerland* |
| 240 003 | *Lübeck* |

# CLASS 290 B-B

**Built:** 1964-74 by MaK/Deutz/Henschel/Jung.
**Engine:** Mercedes 835 Ab/12V 652 TA of 810 kW at 1400 rpm.
**Transmission:** Hydraulic. Voith L206rs.
**Weight:** 78.80 (* 80.00) tonnes.          **Maximum Speed:** 80 (* 70) km/h.
**Length:** 14.320 (* 14.000) m.              **Train Supply:** Not equipped.
**Note:** 290 999 is owned by the German Army.

| | | | | | | | | | |
|---|---|---|---|---|---|---|---|---|---|
| 290 001* | RK | 290 055 | RK | 290 109 | RK | 290 163 | KK | 290 217 | NN2 |
| 290 002* | RK | 290 056 | HH | 290 110 | RM | 290 164 | KK | 290 218 | FK |
| 290 003* | RM | 290 057 | RM | 290 111 | NWH | 290 165 | KK | 290 219 | FK |
| 290 004* | RM | 290 058 | HH | 290 112 | MH1 | 290 166 | KK | 290 220 | KK |
| 290 005* | RM | 290 059 | SSH | 290 113 | MH1 | 290 167 | KK | 290 221 | KK |
| 290 006* | TK | 290 060 | SSH | 290 114 | NWH | 290 168 | MIH | 290 222 | KK |
| 290 007* | RM | 290 061 | FK | 290 115 | RK | 290 169 | KK | 290 223 | KK |
| 290 008* | RM | 290 062 | FK | 290 116 | RHL | 290 170 | KK | 290 224 | KK |
| 290 009* | RM | 290 063 | RK | 290 117 | RHL | 290 171 | EHG | 290 225 | KK |
| 290 010* | RM | 290 064 | EHG | 290 118 | RHL | 290 172 | HH | 290 226 | HBS |
| 290 011* | SSH | 290 065 | EHG | 290 119 | FD | 290 173 | KK | 290 227 | EHM |
| 290 012* | SSH | 290 066 | EHG | 290 120 | FD | 290 174 | KK | 290 228 | TK |
| 290 013* | SSH | 290 067 | RM | 290 121 | NN2 | 290 175 | KK | 290 229 | TK |
| 290 014* | TU | 290 068 | EDO | 290 122 | RM | 290 176 | KK | 290 230 | RHL |
| 290 015* | TU | 290 069 | EHM | 290 123 | RM | 290 177 | MH1 | 290 231 | RHL |
| 290 016* | TK | 290 070 | EHM | 290 124 | RM | 290 178 | NN2 | 290 232 | RHL |
| 290 017* | RM | 290 071 | EHM | 290 125 | MH1 | 290 179 | NN2 | 290 233 | RHL |
| 290 018* | RM | 290 072 | EHM | 290 126 | MH1 | 290 180 | RHL | 290 234 | FF2 |
| 290 019* | TU | 290 073 | EHM | 290 127 | RM | 290 181 | RHL | 290 235 | FF2 |
| 290 020* | TK | 290 074 | EHM | 290 128 | NRH | 290 182 | RHL | 290 236 | FG |
| 290 021 | ESIE | 290 075 | EHG | 290 129 | ESIE | 290 183 | HBS | 290 237 | NN2 |
| 290 022 | EDO | 290 076 | EHG | 290 130 | EDO | 290 184 | TU | 290 238 | NN2 |
| 290 023 | EOB | 290 077 | EHG | 290 131 | SSH | 290 185 | HH | 290 239 | NN2 |
| 290 024 | SSH | 290 078 | EHG | 290 132 | TK | 290 186 | HH | 290 240 | HBS |
| 290 025 | FG | 290 079 | EHG | 290 133 | TK | 290 187 | HH | 290 241 | HH |
| 290 026 | FG | 290 080 | EHG | 290 134 | SSH | 290 188 | MH1 | 290 242 | HBS |
| 290 027 | TK | 290 081 | EHG | 290 135 | RHL | 290 189 | RM | 290 243 | HBS |
| 290 028 | RM | 290 082 | RM | 290 136 | HH | 290 190 | MH1 | 290 244 | HBS |
| 290 029 | MIH | 290 083 | RM | 290 137 | HH | 290 191 | NWH | 290 245 | NN2 |
| 290 030 | ESIE | 290 084 | FF2 | 290 138 | TK | 290 192 | MIH | 290 246 | KK |
| 290 031 | EDO | 290 085 | FF2 | 290 139 | TK | 290 193 | MIH | 290 247 | KK |
| 290 032 | EDO | 290 086 | FF2 | 290 140 | NWH | 290 194 | MH1 | 290 248 | KK |
| 290 033 | EOB | 290 087 | NRH | 290 141 | NN2 | 290 195 | KK | 290 249 | KK |
| 290 034 | EOB | 290 088 | FF2 | 290 142 | FG | 290 196 | KK | 290 250 | NN2 |
| 290 035 | EOB | 290 089 | FF2 | 290 143 | FG | 290 197 | MH1 | 290 251 | KM |
| 290 036 | EOB | 290 090 | FK | 290 144 | FG | 290 198 | MH1 | 290 252 | KM |
| 290 037 | EOB | 290 091 | RM | 290 145 | NN2 | 290 199 | KK | 290 253 | KK |
| 290 038 | EOB | 290 092 | RM | 290 146 | FG | 290 200 | KK | 290 254 | KM |
| 290 039 | ESIE | 290 093 | EDO | 290 147 | EHM | 290 201 | NRH | 290 255 | KM |
| 290 040 | ESIE | 290 094 | ESIE | 290 148 | MH1 | 290 202 | MH1 | 290 256 | KM |
| 290 041 | EHG | 290 095 | EDO | 290 149 | MIH | 290 203 | KK | 290 257 | KM |
| 290 042 | EHG | 290 096 | NN2 | 290 150 | MH1 | 290 204 | KK | 290 258 | KM |
| 290 043 | TU | 290 097 | HH | 290 151 | EHM | 290 205 | MIH | 290 259 | KM |
| 290 044 | TK | 290 098 | HH | 290 152 | EHM | 290 206 | MIH | 290 260 | KM |
| 290 045 | TK | 290 099 | HH | 290 153 | MIH | 290 207 | MIH | 290 261 | HBS |
| 290 046 | FF2 | 290 100 | RHL | 290 154 | MIH | 290 208 | MIH | 290 262 | EHM |
| 290 047 | FF2 | 290 101 | MH1 | 290 155 | MIH | 290 209 | NWH | 290 263 | KK |
| 290 048 | NRH | 290 102 | MH1 | 290 156 | MIH | 290 210 | MH1 | 290 264 | HBS |
| 290 049 | ESIE | 290 103 | MH1 | 290 157 | MIH | 290 211 | FK | 290 265 | ESIE |
| 290 050 | SSH | 290 104 | EHM | 290 158 | HH | 290 212 | FK | 290 266 | EOB |
| 290 051 | SSH | 290 105 | EHM | 290 159 | HBS | 290 213 | RM | 290 267 | EOB |
| 290 052 | EOB | 290 106 | EHM | 290 160 | HBS | 290 214 | SSH | 290 268 | EOB |
| 290 053 | FF2 | 290 107 | EHM | 290 161 | KK | 290 215 | HH | 290 269 | EDO |
| 290 054 | EOB | 290 108 | EHM | 290 162 | KK | 290 216 | NN2 | 290 270 | KM |

# GERMANY

| | | | | | | | | | |
|---|---|---|---|---|---|---|---|---|---|
| 290 271 | EHG | 290 299 | EDO | 290 327 | MIH | 290 355 | EDO | 290 383 | NN2 |
| 290 272 | EHG | 290 300 | EOB | 290 328 | STR | 290 356 | KM | 290 384 | NN2 |
| 290 273 | MIH | 290 301 | EOB | 290 329 | STR | 290 357 | KK | 290 385 | NN2 |
| 290 274 | MIH | 290 302 | EOB | 290 330 | KM | 290 358 | NN2 | 290 386 | NN2 |
| 290 275 | TK | 290 303 | EOB | 290 331 | NN2 | 290 359 | NN2 | 290 387 | NN2 |
| 290 276 | TK | 290 304 | EOB | 290 332 | KM | 290 360 | NN2 | 290 388 | NN2 |
| 290 277 | KK | 290 305 | EOB | 290 333 | TU | 290 361 | NN2 | 290 389 | NWH |
| 290 278 | KK | 290 306 | KM | 290 334 | TU | 290 362 | HBS | 290 390 | KM |
| 290 279 | EOB | 290 307 | KM | 290 335 | TU | 290 363 | HBS | 290 391 | KM |
| 290 280 | EOB | 290 308 | RK | 290 336 | TU | 290 364 | TU | 290 392 | KM |
| 290 281 | EOB | 290 309 | RK | 290 337 | TU | 290 365 | TU | 290 393 | KM |
| 290 282 | EDO | 290 310 | SKL | 290 338 | FK | 290 366 | MH1 | 290 394 | KM |
| 290 283 | NN2 | 290 311 | TK | 290 339 | FD | 290 367 | TU | 290 395 | NN2 |
| 290 284 | EOB | 290 312 | TK | 290 340 | FD | 290 368 | FD | 290 396 | EDO |
| 290 285 | TK | 290 313 | TK | 290 341 | HBS | 290 369 | FD | 290 397 | EOB |
| 290 286 | TK | 290 314 | TK | 290 342 | HH | 290 370 | FD | 290 398 | EOB |
| 290 287 | TK | 290 315 | TK | 290 343 | RK | 290 371 | FD | 290 399 | KM |
| 290 288 | TK | 290 316 | FK | 290 344 | KM | 290 372 | EOB | 290 400 | NRH |
| 290 289 | TK | 290 317 | FK | 290 345 | KM | 290 373 | EOB | 290 401 | NRH |
| 290 290 | MIH | 290 318 | FK | 290 346 | KK | 290 374 | EHM | 290 402 | NRH |
| 290 291 | NN2 | 290 319 | FK | 290 347 | KK | 290 375 | EOB | 290 403 | NN2 |
| 290 292 | NN2 | 290 320 | RK | 290 348 | KK | 290 376 | EOB | 290 404 | NN2 |
| 290 293 | NN2 | 290 321 | RK | 290 349 | NWH | 290 377 | KM | 290 405 | NN2 |
| 290 294 | NWH | 290 322 | RK | 290 350 | KM | 290 378 | KM | 290 406 | NN2 |
| 290 295 | NWH | 290 323 | TU | 290 351 | KM | 290 379 | EOB | 290 407 | NN2 |
| 290 296 | EOB | 290 324 | TU | 290 352 | EDO | 290 380 | EOB | 290 999 | KK |
| 290 297 | EOB | 290 325 | TU | 290 353 | EDO | 290 381 | EOB | | |
| 290 298 | EDO | 290 326 | MIH | 290 354 | EOB | 290 382 | NN2 | | |

## CLASS 291     B-B

**Built:** 1974-78 (*1965) by MaK/Jung.
**Engine:** MaK 8 M 282 Ak of 810 kW at 927 rpm.
**Transmission:** Hydraulic. Voith L206rsb.
**Weight:** 80.00 tonnes.     **Maximum Speed:** 80 km/h.
**Length:** 14.320 m.     **Train Supply:** Not equipped.

| | | | | | | | | | |
|---|---|---|---|---|---|---|---|---|---|
| 291 001 | AH4 | 291 022 | AH4 | 291 043 | AH4 | 291 064 | HB | 291 084 | HBH |
| 291 002 | AH4 | 291 023 | AH4 | 291 044 | AH4 | 291 065 | HB | 291 085 | HBH |
| 291 003 | AH4 | 291 024 | AH4 | 291 045 | HO | 291 066 | HB | 291 086 | HBH |
| 291 004 | AH4 | 291 025 | AH4 | 291 046 | HO | 291 067 | HB | 291 087 | HBH |
| 291 005 | AH4 | 291 026 | AH4 | 291 047 | HBH | 291 068 | HB | 291 088 | HBH |
| 291 006 | AH4 | 291 027 | AH4 | 291 048 | HBH | 291 069 | HB | 291 089 | HBH |
| 291 007 | AH4 | 291 028 | AH4 | 291 049 | HBH | 291 070 | HB | 291 090 | HE |
| 291 008 | AH4 | 291 029 | AH4 | 291 050 | HBH | 291 071 | HB | 291 091 | HO |
| 291 009 | AH4 | 291 030 | AH4 | 291 051 | HB | 291 072 | HO | 291 092 | HE |
| 291 010 | AH4 | 291 031 | AH4 | 291 052 | HB | 291 073 | HB | 291 093 | HBH |
| 291 011 | AH4 | 291 032 | AH4 | 291 053 | HB | 291 074 | HB | 291 094 | HO |
| 291 012 | AH4 | 291 033 | AH4 | 291 054 | HB | 291 075 | HB | 291 095 | HO |
| 291 013 | AH4 | 291 034 | AH4 | 291 055 | HB | 291 076 | HB | 291 096 | HO |
| 291 014 | AH4 | 291 035 | AH4 | 291 056 | HB | 291 077 | HB | 291 097 | HO |
| 291 015 | AH4 | 291 036 | AH4 | 291 057 | HB | 291 078 | HB | 291 098 | HO |
| 291 016 | AH4 | 291 037 | AH4 | 291 058 | HB | 291 079 | HB | 291 099 | HO |
| 291 017 | AH4 | 291 038 | AH4 | 291 059 | HB | 291 080 | HB | 291 100 | HO |
| 291 018 | AH4 | 291 039 | AH4 | 291 060 | HB | 291 081 | HO | 291 901* | HB |
| 291 019 | AH4 | 291 040 | AH4 | 291 061 | HB | 291 082 | HB | 291 902* | HB |
| 291 020 | AH4 | 291 041 | AH4 | 291 062 | HB | 291 083 | HBH | 291 903* | HB |
| 291 021 | AH4 | 291 042 | AH4 | 291 063 | HB | | | | |

## CLASS 293     B-B

**Built:** 1982 by LEW.
**Engine:** Johannisthal 12 KVD 21 AL3 of 736 kW at 1500 rpm.
**Transmission:** Hydraulic. Strömungsmaschinen GSR 30/5.7 AQ.

**Weight:** 62.20 tonnes.
**Length:** 14.240 m.
**Note:** Previously DR Class 111.
**Maximum Speed:** 65 km/h
**Train Supply:** Not equipped.

| | | | |
|---|---|---|---|
| 293 107 | LW | 293 128 | LW |

## CLASS 298        B-B

**Built:** 1964-82 by LEW as Classes 201 and 293. Rebuilt 1978 by DR.
**Engine:** Johannisthal 12 KVD 18/21 A2 of 736 kW at 1500 rpm.
**Transmission:** Hydraulic. Strömungsmachinen GS 20-20/4.1.
**Weight:** 63.70 tonnes.
**Length:** 14.240 m.
**Note:** Previously DR Class 108.
**Maximum Speed:** 60 km/h.
**Train Supply:** Not equipped.

| | | | | | | | | | |
|---|---|---|---|---|---|---|---|---|---|
| 298 044 | DH | 298 072 | DZW | 298 122 | DC | 298 306 | BFG | 298 322 | LL1 |
| 298 045 | LH1 | 298 074 | BFG | 298 124 | BWUR | 298 307 | WR | 298 323 | LL1 |
| 298 046 | LH1 | 298 079 | DH | 298 129 | WSR | 298 308 | BWUR | 298 324 | LL1 |
| 298 047 | DZW | 298 080 | DZW | 298 139 | DH | 298 309 | DH | 298 325 | LMR |
| 298 048 | DH | 298 081 | DH | 298 142 | DH | 298 310 | WR | 298 326 | LL1 |
| 298 050 | DH | 298 084 | WSR | 298 150 | BWUR | 298 311 | DC | 298 327 | UE |
| 298 051 | DH | 298 085 | DC | 298 151 | DH | 298 312 | WR | 298 328 | LL1 |
| 298 052 | BWUR | 298 086 | BWUR | 298 155 | BWUR | 298 313 | LMR | 298 329 | LMR |
| 298 054 | WSR | 298 088 | DZW | 298 156 | BCS | 298 314 | DH | 298 330 | LL1 |
| 298 055 | BWUR | 298 091 | DZW | 298 161 | BCS | 298 315 | LH1 | 298 331 | BFG |
| 298 058 | DZW | 298 094 | DZW | 298 163 | DZW | 298 316 | WR | 298 332 | LL1 |
| 298 060 | BWUR | 298 099 | LMR | 298 301 | LMR | 298 317 | BFG | 298 333 | LL1 |
| 298 062 | WSR | 298 100 | UE | 298 302 | BCS | 298 318 | WR | 298 334 | LMR |
| 298 065 | UE | 298 102 | BFG | 298 303 | DZW | 298 319 | LMR | 298 335 | LMR |
| 298 069 | DZW | 298 104 | DZW | 298 304 | DC | 298 320 | WR | 298 336 | LMR |
| 298 071 | DZW | 298 110 | DZW | 298 305 | LMR | 298 321 | LL1 | 298 337 | LH1 |

## STANDARD GAUGE SHUNTING LOCOMOTIVES

## CLASS 310.1        B

**Built:** 1932-50 by Schwartzkopff/Henschel/Borsig/Jung/Krauss-Maffei/Deutz/O & K/Gmeinder/Windhoff/LKM.
**Engine:** Johannisthal 6 KVD 14.5 SRW of 92 kW at 1250 rpm.
**Transmission:** Mechanical.
**Weight:** 16.00 tonnes.
**Length:** 6.450 m.
**Maximum Speed:** 30 km/h.
**Note:** Previously DR Class 100.

| | | | | | | | | | |
|---|---|---|---|---|---|---|---|---|---|
| 310 106 | (Z) | 310 275 | (Z) | 310 429 | US | 310 528 | (Z) | 310 732 | (Z) |
| 310 107 | US | 310 279 | (Z) | 310 432 | US | 310 543 | US | 310 733 | UEI |
| 310 110 | BSC | 310 291 | LLW | 310 434 | (Z) | 310 589 | UN | 310 734 | US |
| 310 116 | UEI | 310 295 | UN | 310 435 | UEI | 310 617 | LHB | 310 740 | (Z) |
| 310 123 | LLW | 310 352 | BHF | 310 436 | (Z) | 310 625 | BHF | 310 742 | LLW |
| 310 127 | (Z) | 310 356 | (Z) | 310 439 | US | 310 626 | LM | 310 751 | LLW |
| 310 201 | (Z) | 310 357 | LM | 310 440 | US | 310 631 | BPKR | 310 762 | (Z) |
| 310 206 | LLW | 310 406 | (Z) | 310 441 | UN | 310 645 | UG | 310 763 | BPKR |
| 310 208 | LLW | 310 407 | (Z) | 310 442 | UW | 310 704 | (Z) | 310 764 | (Z) |
| 310 209 | (Z) | 310 424 | US | 310 443 | US | 310 705 | US | 310 768 | (Z) |
| 310 211 | DC | 310 425 | (Z) | 310 457 | (Z) | 310 708 | BPKR | 310 770 | (Z) |
| 310 216 | LLW | 310 426 | UEI | 310 494 | (Z) | 310 714 | UW | 310 771 | (Z) |
| 310 227 | LLW | 310 427 | (Z) | 310 507 | (Z) | 310 723 | (Z) | 310 772 | UW |
| 310 247 | (Z) | 310 428 | (Z) | 310 517 | LLW | 310 731 | LLW | 310 773 | (Z) |
| 310 273 | LLW | | | | | | | | |

## CLASS 310.8        B

**Built:** 1932-50 by Schwartzkopff/Henschel/Borsig/Jung/Krauss-Maffei/Deutz/O & K/Gmeinder/Windhoff/LKM.
**Engine:** Johannisthal 6 KVD 14.5 SRW of 92 kW at 1250 rpm.
**Transmission:** Hydraulic.
**Length:** 6.450 m.

# GERMANY
**Weight:** 17.00 tonnes.　　　　　　　　**Maximum Speed:** 30 Km/h.
**Note:** Previously DR Class 100.

| | | | | | | | | | |
|---|---|---|---|---|---|---|---|---|---|
| 310 800 | DZW | 310 821 | UN | 310 858 | LLW | 310 910 | (Z) | 310 941 | LLW |
| 310 801 | (Z) | 310 822 | (Z) | 310 859 | LLW | 310 911 | LLW | 310 942 | LLW |
| 310 806 | (Z) | 310 827 | (Z) | 310 860 | UW | 310 912 | UEI | 310 944 | UN |
| 310 807 | LHB | 310 830 | (Z) | 310 874 | LLW | 310 918 | BPKR | 310 947 | BSE |
| 310 808 | (Z) | 310 836 | LLW | 310 877 | LLW | 310 924 | LLW | 310 948 | (Z) |
| 310 810 | LLW | 310 838 | (Z) | 310 879 | LLW | 310 926 | (Z) | 310 949 | (Z) |
| 310 811 | (Z) | 310 840 | (Z) | 310 893 | UEI | 310 935 | BPKR | 310 951 | LLW |
| 310 812 | BSC | 310 842 | (Z) | 310 897 | BSE | 310 937 | (Z) | 310 954 | (Z) |
| 310 818 | (Z) | 310 848 | (Z) | | | | | | |

## CLASS 311.1　　　　　　　　　　　　　　　　　　B

**Built:** 1960-63 by LKM.
**Engine:** Johannisthal 6 KVD 18 SRW of 132 kW at 1500 rpm.
**Transmission:** Hydraulic. Strömungsmachinen GSR 12/3.7.
**Weight:** 21.50 tonnes.　　　　　　　　**Maximum Speed:** 37 km/h.
**Length:** 6.940 m.　　　　　　　　**Note:** Previously DR Class 101.

| | | | | | |
|---|---|---|---|---|---|
| 311 114 | LH1 | 311 115 | (Z) | 311 123 | (Z) |

## CLASS 311.5　　　　　　　　　　　　　　　　　　B

**Built:** 1960-63 by LKM. Rebuilt 1975-79 by DR Halle Works.
**Engine:** Johannisthal 6 KVD 18/15-1 SRW of 162 kW at 1510 rpm.
**Transmission:** Hydraulic. Strömungsmachinen GSU 20/4.2.
**Weight:** 21.50 tonnes.　　　　　　　　**Maximum Speed:** 42 km/h.
**Length:** 6.940 m.　　　　　　　　**Note:** Previously DR Class 101.

| | | | | | | | | | |
|---|---|---|---|---|---|---|---|---|---|
| 311 501 | LL1 | 311 543 | LL1 | 311 588 | (Z) | 311 633 | (Z) | 311 690 | LLW |
| 311 504 | LL1 | 311 546 | BHW | 311 589 | (Z) | 311 638 | (Z) | 311 691 | US |
| 311 505 | LS | 311 552 | LL1 | 311 591 | BWUR | 311 644 | LL1 | 311 692 | (Z) |
| 311 507 | (Z) | 311 556 | LLW | 311 594 | LL1 | 311 650 | (Z) | 311 693 | (Z) |
| 311 513 | LLW | 311 558 | (Z) | 311 599 | BPKR | 311 658 | BPKR | 311 696 | (Z) |
| 311 515 | LLW | 311 560 | (Z) | 311 600 | LM | 311 659 | (Z) | 311 697 | WNT |
| 311 517 | US | 311 561 | (Z) | 311 601 | LL1 | 311 660 | (Z) | 311 699 | UEI |
| 311 518 | LL1 | 311 562 | LH1 | 311 604 | WS | 311 661 | LM | 311 701 | LH1 |
| 311 519 | WR | 311 563 | US | 311 606 | LLW | 311 663 | (Z) | 311 702 | (Z) |
| 311 520 | UW | 311 564 | LL1 | 311 608 | WP | 311 664 | LH1 | 311 703 | (Z) |
| 311 521 | (Z) | 311 567 | UM | 311 609 | US | 311 668 | LL1 | 311 708 | BPKR |
| 311 522 | (Z) | 311 568 | (Z) | 311 611 | (Z) | 311 674 | BPKR | 311 709 | WNT |
| 311 523 | BWUR | 311 572 | (Z) | 311 614 | WS | 311 677 | LL1 | 311 712 | LLW |
| 311 524 | (Z) | 311 574 | (Z) | 311 616 | LL1 | 311 678 | UG | 311 713 | WNT |
| 311 530 | WR | 311 576 | UM | 311 619 | BWUR | 311 680 | (Z) | 311 714 | (Z) |
| 311 531 | (Z) | 311 577 | (Z) | 311 620 | BPKR | 311 682 | LHB | 311 715 | US |
| 311 532 | LL1 | 311 579 | LM | 311 621 | (Z) | 311 683 | BPKR | 311 718 | (Z) |
| 311 533 | UW | 311 580 | UM | 311 626 | UM | 311 684 | (Z) | 311 720 | (Z) |
| 311 540 | (Z) | 311 581 | LH1 | 311 627 | BPKR | 311 686 | (Z) | 311 724 | BPKR |
| 311 541 | LL1 | 311 582 | (Z) | 311 628 | (Z) | 311 687 | (Z) | 311 725 | (Z) |
| 311 542 | LLW | 311 585 | LL1 | 311 631 | LS | 311 688 | US | | |

## CLASS 312.0　　　　　　　　　　　　　　　　　　B

**Built:** 1968-70 by LKM.
**Engine:** Roßlau 6 KVD 18/15-1 SRW of 162 kW at 1510 rpm.
**Transmission:** Hydraulic. Strömungsmachinen GSU 20/4.5.
**Weight:** 24.00 tonnes.　　　　　　　　**Maximum Speed:** 35 km/h.
**Length:** 6.940 m.　　　　　　　　**Note:** Previously DR Class 102.

| | | | | | | | | | |
|---|---|---|---|---|---|---|---|---|---|
| 312 001 | LHB | 312 011 | DH | 312 018 | DG | 312 023 | DH | 312 035 | DR |
| 312 004 | LW | 312 012 | DR | 312 019 | LL1 | 312 024 | DRC | 312 038 | LH1 |
| 312 007 | BHW | 312 013 | LH1 | 312 020 | LL2 | 312 025 | DR | 312 039 | LL1 |
| 312 008 | DG | 312 017 | LH1 | 312 021 | BSC | 312 029 | LL2 | 312 040 | (Z) |

| 312 041 | BSE | 312 045 | LM | 312 055 | WR | 312 060 | (Z) | 312 066 | (Z) |
|---------|-----|---------|-----|---------|-----|---------|-----|---------|-----|
| 312 043 | LM | 312 050 | UE | 312 057 | DH | 312 062 | (Z) | 312 067 | UEI |
| 312 044 | LS | 312 051 | (Z) | 312 058 | (Z) | 312 065 | (Z) | 312 081 | (Z) |

# CLASS 312.1 B

**Built:** 1970-71 by LKM.
**Engine:** Roßlau 6 VD 18/15-1 SRW of 162 kw at 1510 rpm.
**Transmission:** Hydraulic. Strömungsmachinen GSU 20/4.5.
**Weight:** 24.30 tonnes. **Maximum Speed:** 40 km/h.
**Length:** 8.000 m. **Note:** Previously DR Class 102.

| 312 101 | UE | 312 131 | DZW | 312 164 | (Z) | 312 197 | BSC | 312 229 | LLW |
|---------|-----|---------|-----|---------|-----|---------|-----|---------|-----|
| 312 102 | LM | 312 132 | DZW | 312 165 | (Z) | 312 198 | BHW | 312 230 | LS |
| 312 103 | LM | 312 133 | (Z) | 312 167 | LHB | 312 199 | BSC | 312 231 | BSE |
| 312 104 | LM | 312 134 | BPKR | 312 168 | BSC | 312 200 | LL1 | 312 232 | BSE |
| 312 105 | LH1 | 312 135 | BPKR | 312 169 | BHW | 312 201 | LL1 | 312 233 | BSE |
| 312 106 | LM | 312 137 | (Z) | 312 171 | BSC | 312 202 | BPKR | 312 234 | BFG |
| 312 107 | LMR | 312 138 | BFG | 312 172 | DZW | 312 203 | DG | 312 235 | BFG |
| 312 109 | WNT | 312 139 | BPKR | 312 173 | DG | 312 204 | DH | 312 236 | LHB |
| 312 110 | WSR | 312 140 | DZW | 312 174 | BPKR | 312 205 | BSC | 312 237 | LL1 |
| 312 111 | WNT | 312 141 | WR | 312 175 | BPKR | 312 207 | UG | 312 238 | WR |
| 312 112 | WR | 312 142 | BSC | 312 176 | DG | 312 209 | BSE | 312 239 | BPKR |
| 312 113 | UE | 312 143 | LW | 312 178 | WW | 312 210 | UEI | 312 240 | BPKR |
| 312 114 | LHB | 312 144 | LH1 | 312 179 | WR | 312 211 | UEI | 312 241 | BPKR |
| 312 115 | (Z) | 312 145 | BWUR | 312 181 | (Z) | 312 212 | LM | 312 242 | (Z) |
| 312 116 | (Z) | 312 147 | BSC | 312 183 | DZW | 312 213 | LM | 312 243 | LLW |
| 312 117 | UN | 312 152 | LS | 312 184 | BSE | 312 215 | WR | 312 244 | LL1 |
| 312 118 | UN | 312 153 | WS | 312 185 | BWUR | 312 217 | WR | 312 245 | LH1 |
| 312 119 | (Z) | 312 154 | WS | 312 186 | BFG | 312 218 | (Z) | 312 246 | DC |
| 312 120 | LM | 312 155 | WNT | 312 187 | DC | 312 221 | WS | 312 247 | DC |
| 312 122 | LS | 312 156 | UN | 312 188 | (Z) | 312 222 | BFG | 312 248 | (Z) |
| 312 123 | UEI | 312 157 | UEI | 312 190 | LL1 | 312 223 | BSE | 312 249 | (Z) |
| 312 124 | WW | 312 159 | UE | 312 191 | LL1 | 312 224 | BPKR | 312 252 | (Z) |
| 312 125 | (Z) | 312 160 | LM | 312 193 | UN | 312 225 | WR | 312 253 | DZW |
| 312 126 | WNT | 312 161 | US | 312 194 | UE | 312 226 | WP | 312 255 | DG |
| 312 127 | WP | 312 162 | UN | 312 195 | UN | 312 227 | WSR | 312 256 | LL1 |
| 312 129 | WNT | 312 163 | BPKR | 312 196 | (Z) | 312 228 | LM | 312 257 | LHB |
| 312 130 | DZW | | | | | | | | |

# CLASS 323/324 B

**Built:** 1944-65 by Schwartzkopff/Henschel/Borsig/Jung/Krauss-Maffei/Deutz/O & K/Gmeinder/Windhoff.
**Engine:** Kaelble GN 130s of 94 kW at 1260 rpm or Deutz A6M 517 of 87 kW at 1300 rpm or Deutz A6M 617 of 94 kW (128 hp) at 1300 rpm.
**Transmission:** Hydraulic. Voith L33U. **Maximum Speed:** 45 km/h.
**Weight:** 17.00 tonnes. **Length:** 6.450 m.

| 323 072 | (Z) | 323 184 | (Z) | 323 264 | HO | 323 331 | HO | 323 617 | KG |
|---------|-----|---------|-----|---------|-----|---------|-----|---------|-----|
| 323 079 | RF | 323 186 | AH4 | 323 270 | AH4 | 323 332 | KG | 323 641 | HO |
| 323 083 | KG | 323 188 | AH4 | 323 273 | EOB | 323 339 | EHM | 323 658 | RK |
| 323 088 | HBS | 323 190 | AH4 | 323 276 | EHG | 323 341 | AH4 | 323 662 | FD |
| 323 099 | HB | 323 192 | EHG | 323 279 | KG | 323 342 | MA | 323 680 | (Z) |
| 323 101 | HB | 323 208 | RK | 323 287 | HO | 323 352 | HB | 323 691 | FF2 |
| 323 113 | HH | 323 214 | KG | 323 288 | (Z) | 323 440 | NRH | 323 710 | HBS |
| 323 128 | RK | 323 216 | (Z) | 323 291 | HO | 323 460 | RO | 323 715 | FD |
| 323 139 | (Z) | 323 224 | (Z) | 323 302 | HH | 323 465 | (Z) | 323 716 | FD |
| 323 145 | NN2 | 323 228 | (Z) | 323 303 | HH | 323 522 | EDO | 323 729 | TK |
| 323 157 | KM | 323 229 | (Z) | 323 305 | KG | 323 546 | KG | 323 730 | NHO |
| 323 161 | AH4 | 323 239 | (Z) | 323 322 | AH4 | 323 549 | TK | 323 734 | RO |
| 323 165 | AH4 | 323 241 | (Z) | 323 323 | HH | 323 556 | (Z) | 323 737 | TK |
| 323 169 | KM | 323 254 | HB | 323 325 | KG | 323 576 | (Z) | 323 738 | (Z) |
| 323 173 | EDO | 323 257 | AH4 | 323 327 | TK | 323 582 | FG | 323 743 | (Z) |
| 323 178 | (Z) | 323 260 | HHX | 323 330 | RHL | 323 587 | KG | 323 747 | FF2 |

# GERMANY

| | | | | |
|---|---|---|---|---|
| 323 749 FF2 | 323 790 RHL | 323 821 (Z) | 323 849 (Z) | 323 870 HG |
| 323 756 NN2 | 323 797 ESIE | 323 822 NWH | 323 852 KG | 323 871 NHO |
| 323 757 NLF | 323 798 FD | 323 827 FD | 323 856 FK | 323 873 HH |
| 323 758 NN2 | 323 800 RO | 323 832 RM | 323 859 RM | 323 880 RK |
| 323 760 NWH | 323 802 MH1 | 323 835 RM | 323 860 (Z) | 323 881 (Z) |
| 323 763 NN2 | 323 804 RHL | 323 838 RHL | 323 862 KG | 323 998 (Z) |
| 323 770 HBS | 323 816 RM | 323 839 RO | 323 863 (Z) | 324 023 HB |
| 323 777 RK | 323 818 FD | 323 842 FG | 323 867 HBS | 324 052 FG |
| 323 780 HH | 323 819 FF2 | 323 847 FK | | |

# CLASS 332                                                                  B

**Built:** 1959-66 by Gmeinder/Jung/O & K.
**Engine:** MWM RHS 518 A of 177 kW at 1600 rpm.
**Transmission:** Hydraulic. Voith L213U.    **Maximum Speed:** 45 km/h.
**Weight:** 22.30 tonnes.    **Length:** 7.830 m.

| | | | | |
|---|---|---|---|---|
| 332 002 HG | 332 051 NRH | 332 108 KM | 332 160 TU | 332 215 SKL |
| 332 005 KG | 332 052 NRH | 332 109 KA | 332 161 (Z) | 332 216 STR |
| 332 006 RO | 332 053 MMF | 332 110 EOB | 332 163 (Z) | 332 217 NRH |
| 332 007 AH4 | 332 054 MH1 | 332 111 KG | 332 164 (Z) | 332 218 FG |
| 332 008 HG | 332 057 TU | 332 112 KM | 332 165 RM | 332 219 FF2 |
| 332 009 FK | 332 058 (Z) | 332 113 RO | 332 166 FD | 332 220 RM |
| 332 010 HH | 332 060 MH1 | 332 114 RM | 332 167 HB | 332 221 (Z) |
| 332 011 EOB | 332 061 NN2 | 332 115 RF | 332 168 HE | 332 222 MMF |
| 332 012 EHM | 332 062 NRH | 332 119 HO | 332 169 (Z) | 332 223 MA |
| 332 013 KG | 332 064 FF2 | 332 120 HH | 332 170 (Z) | 332 224 RK |
| 332 014 HO | 332 065 RO | 332 121 AH4 | 332 171 STR | 332 226 (Z) |
| 332 015 (Z) | 332 066 NWH | 332 122 AH4 | 332 172 RO | 332 227 FG |
| 332 016 KA | 332 067 NN2 | 332 123 AH4 | 332 173 FK | 332 228 MH1 |
| 332 017 (Z) | 332 068 (Z) | 332 125 FF2 | 332 174 FF2 | 332 229 FK |
| 332 018 KM | 332 069 MH1 | 332 126 (Z) | 332 175 FD | 332 230 RM |
| 332 019 AFL | 332 070 MKP | 332 127 KM | 332 176 FG | 332 231 TT |
| 332 020 SKL | 332 073 EOB | 332 128 RK | 332 177 FD | 332 232 TK |
| 332 021 RSI | 332 074 (Z) | 332 129 KG | 332 182 EHG | 332 233 FG |
| 332 022 TU | 332 075 NN2 | 332 130 ESIE | 332 183 EOB | 332 234 MA |
| 332 023 RM | 332 076 NN2 | 332 131 HH | 332 184 HH | 332 235 MIH |
| 332 024 RO | 332 077 KG | 332 132 (Z) | 332 185 (Z) | 332 236 RSI |
| 332 025 MIH | 332 080 FD | 332 134 AFL | 332 186 HG | 332 238 FF2 |
| 332 026 NWH | 332 081 AH4 | 332 135 (Z) | 332 187 KA | 332 239 MMF |
| 332 027 NLF | 332 082 (Z) | 332 136 HO | 332 188 KA | 332 240 KM |
| 332 028 HG | 332 083 HB | 332 137 KM | 332 189 HBS | 332 241 HB |
| 332 029 FK | 332 084 HB | 332 138 EHG | 332 190 HO | 332 242 KM |
| 332 030 RM | 332 085 (Z) | 332 139 KM | 332 193 FG | 332 243 MA |
| 332 031 FG | 332 087 (Z) | 332 140 EOB | 332 194 (Z) | 332 244 HO |
| 332 032 FG | 332 088 HE | 332 141 KG | 332 195 FG | 332 245 NHO |
| 332 033 FG | 332 089 (Z) | 332 142 (Z) | 332 196 KG | 332 246 TU |
| 332 034 RO | 332 090 AFL | 332 143 HB | 332 197 MMF | 332 247 TU |
| 332 035 RSI | 332 091 EOB | 332 144 HBS | 332 198 MA | 332 248 AH4 |
| 332 036 SKL | 332 092 (Z) | 332 145 EDO | 332 199 MH1 | 332 249 AFL |
| 332 037 FF2 | 332 093 (Z) | 332 146 EHM | 332 200 EOB | 332 250 TU |
| 332 038 RHL | 332 094 (Z) | 332 147 HB | 332 201 (Z) | 332 251 TU |
| 332 039 SSH | 332 095 HBS | 332 148 EHM | 332 202 MA | 332 252 HB |
| 332 040 RK | 332 096 KG | 332 149 EOB | 332 203 (Z) | 332 253 HB |
| 332 041 RK | 332 098 KG | 332 150 EOB | 332 204 ESIE | 332 254 HO |
| 332 042 FD | 332 099 HG | 332 151 EHM | 332 205 FG | 332 255 TT |
| 332 043 SSH | 332 100 (Z) | 332 152 HO | 332 206 AH4 | 332 256 TK |
| 332 044 FK | 332 101 HB | 332 153 EOB | 332 207 NN2 | 332 257 EHM |
| 332 045 FG | 332 102 RHL | 332 154 AH4 | 332 208 (Z) | 332 258 MH1 |
| 332 046 FF2 | 332 103 HO | 332 155 (Z) | 332 209 MMF | 332 259 (Z) |
| 332 047 RO | 332 104 HH | 332 156 HH | 332 212 NRH | 332 260 KM |
| 332 048 RSI | 332 105 HO | 332 157 EHG | 332 213 SKL | 332 261 HO |
| 332 049 MH1 | 332 106 EDO | 332 158 STR | 332 214 FG | 332 262 NN2 |
| 332 050 RF | 332 107 EDO | 332 159 TH | | 332 263 NN2 |

| | | | | | | | | |
|---|---|---|---|---|---|---|---|---|---|
| 332 264 | EHM | 332 276 | RSI | 332 288 | HH | 332 301 | NHO | 332 314 | FF2 |
| 332 265 | AH4 | 332 277 | FG | 332 289 | HBS | 332 302 | SSH | 332 315 | FG |
| 332 266 | KG | 332 278 | RM | 332 290 | KG | 332 303 | STR | 332 316 | FG |
| 332 267 | RM | 332 279 | FK | 332 291 | KG | 332 304 | FF2 | 332 317 | FF2 |
| 332 268 | AH4 | 332 280 | KM | 332 292 | HB | 332 305 | FF2 | 332 601 | MH1 |
| 332 269 | AL | 332 281 | EDO | 332 293 | HB | 332 307 | EHM | 332 602 | (Z) |
| 332 270 | NWH | 332 282 | HB | 332 294 | EHM | 332 308 | AH4 | 332 701 | (Z) |
| 332 271 | NN2 | 332 283 | AL | 332 295 | EOB | 332 309 | (Z) | 332 702 | HO |
| 332 272 | AL | 332 284 | EDO | 332 297 | RF | 332 310 | KG | 332 801 | AFL |
| 332 273 | TU | 332 285 | KM | 332 298 | TT | 332 311 | SSH | 332 901 | NRH |
| 332 274 | MA | 332 286 | AL | 332 299 | TU | 332 312 | FF2 | 332 902 | (Z) |
| 332 275 | RO | 332 287 | AL | 332 300 | TH | 332 313 | SSH | | |

## CLASS 333                                                                    B

**Built:** 1965-78 by Gmeinder/Jung/O & K.
**Engine:** MWM RHS 518 A of 177 kW at 1600 rpm.
**Transmission:** Hydraulic. Voith L203ku (* L303).
**Weight:** 22.30 tonnes.                    **Maximum Speed:** 45 km/h.
**Length:** 7.830 m.

| | | | | | | | | |
|---|---|---|---|---|---|---|---|---|---|
| 333 001* | NN2 | 333 025 | KM | 333 048 | AFL | 333 063 | NLF | 333 075 | FF2 |
| 333 002 | AFL | 333 026 | EHG | 333 050 | AH4 | 333 064 | NN2 | 333 076 | FF2 |
| 333 003 | AFL | 333 028 | AOP | 333 051 | HO | 333 065 | NHO | 333 077 | NN2 |
| 333 005 | AL | 333 031 | MH1 | 333 053 | TH | 333 067 | MMF | 333 080 | KM |
| 333 008 | EOB | 333 040 | AH4 | 333 054 | NLF | 333 068 | TH | 333 081 | FF2 |
| 333 012 | AH4 | 333 041 | EHM | 333 055 | NN2 | 333 069 | TH | 333 093 | TH |
| 333 015 | EHM | 333 042 | HO | 333 056 | MIH | 333 070 | EHG | 333 094 | NLF |
| 333 016 | AH4 | 333 043 | AH4 | 333 057 | NWH | 333 071 | NRH | 333 096 | AL |
| 333 019 | KM | 333 044 | MH1 | 333 058 | MA | 333 072 | FK | 333 104 | AOP |
| 333 021 | KG | 333 045 | HH | 333 059 | MH1 | 333 073 | FG | 333 135 | FF2 |
| 333 022 | AFL | 333 046 | EHM | 333 062 | NLF | 333 074 | FG | 333 172 | RO |
| 333 024 | KM | 333 047 | AFL | | | | | | |

## CLASS 335                                                                    B

**Built:** 1965-78 by Gmeinder/Jung/O & K.
**Engine:** MWM RHS 518 A (* MWM D 601-6) of 177 kW at 1600 rpm.
**Transmission:** Hydraulic. Voith L203ku.          **Length:** 7.830 m.
**Weight:** 22.30 tonnes.                    **Maximum Speed:** 45 km/h.

| | | | | | | | | |
|---|---|---|---|---|---|---|---|---|---|
| 335 004 | RHL | 335 060 | TU | 335 103 | SKL | 335 127 | RM | 335 152 | TK |
| 335 006 | NN2 | 335 061 | TU | 335 105 | RO | 335 128 | EDO | 335 153 | FF2 |
| 335 007 | TK | 335 066 | MIH | 335 106 | FF2 | 335 129 | SKL | 335 154 | MH1 |
| 335 009 | TK | 335 078 | FF2 | 335 107 | MKP | 335 130 | SKL | 335 155 | FF2 |
| 335 010 | TK | 335 079 | SKL | 335 108 | RM | 335 131 | TK | 335 156 | SSH |
| 335 011 | TK | 335 082 | EOB | 335 109 | TU | 335 132 | MMF | 335 157 | STR |
| 335 013 | KM | 335 083 | HH | 335 110 | TU | 335 133 | MMF | 335 158 | FF2 |
| 335 014 | FG | 335 084 | MIH | 335 111 | TU | 335 134 | TU | 335 159 | SSH |
| 335 017 | RM | 335 085 | EHM | 335 112 | NN2 | 335 136 | STR | 335 160 | FF2 |
| 335 018 | RSI | 335 086 | KM | 335 113 | MKP | 335 137 | EHM | 335 161 | FK |
| 335 023 | SKL | 335 087 | FG | 335 114 | MIH | 335 138 | HBS | 335 162 | MH1 |
| 335 027 | MIH | 335 088 | MMF | 335 115 | MKP | 335 139 | HH | 335 163 | STR |
| 335 029 | SSH | 335 089 | MKP | 335 116 | MKP | 335 140 | RK | 335 164 | NN2 |
| 335 030 | TK | 335 090 | MH1 | 335 117 | MIH | 335 141 | MMF | 335 165 | NN2 |
| 335 032 | MMF | 335 091 | SKL | 335 118 | MMF | 335 142 | SKL | 335 166 | NRH |
| 335 033 | MKP | 335 092 | TU | 335 119 | FG | 335 143 | MMF | 335 167 | MKP |
| 335 034 | MMF | 335 095 | HH | 335 120 | TK | 335 144 | RHL | 335 168 | MMF |
| 335 035 | FK | 335 097 | EHM | 335 121 | TK | 335 146 | RK | 335 169 | MMF |
| 335 036 | STR | 335 098 | HB | 335 122 | NRH | 335 147 | FF2 | 335 170 | RO |
| 335 037 | NRH | 335 099 | HO | 335 123 | TK | 335 148 | TU | 335 171 | RK |
| 335 038 | SSH | 335 100 | EDO | 335 124 | HBS | 335 149 | TU | 335 173 | NN2 |
| 335 039 | SKL | 335 101 | HB | 335 125 | TK | 335 150 | STR | 335 174 | MIH |
| 335 052 | TU | 335 102 | TU | 335 126 | HB | 335 151 | RO | 335 175 | NN2 |

# GERMANY

| | | | | | | | | | |
|---|---|---|---|---|---|---|---|---|---|
| 335 176 | NN2 | 335 192* | HBS | 335 207* | KM | 335 222* | AL | 335 237* | HB |
| 335 177 | NN2 | 335 193* | HH | 335 208* | HH | 335 223* | AL | 335 238* | HH |
| 335 178 | MMF | 335 194* | HB | 335 209* | HH | 335 224* | KM | 335 239* | KM |
| 335 179 | NN2 | 335 195* | EDO | 335 210* | HBS | 335 225* | AH4 | 335 240* | HH |
| 335 180 | FK | 335 196* | EDO | 335 211* | KM | 335 226* | AH4 | 335 241* | HH |
| 335 181 | FF2 | 335 197* | EOB | 335 212* | KM | 335 227* | AH4 | 335 242* | EDO |
| 335 182* | HG | 335 198* | FK | 335 213* | AL | 335 228* | AH4 | 335 243* | FG |
| 335 183* | EDO | 335 199* | FF2 | 335 214* | FG | 335 229* | HO | 335 244* | HB |
| 335 184* | HB | 335 200* | FK | 335 215* | AL | 335 230* | EHM | 335 245* | HBS |
| 335 185* | AL | 335 201* | HH | 335 216* | EOB | 335 231* | HBS | 335 246* | HE |
| 335 186* | HG | 335 202* | EHM | 335 217* | EHM | 335 232* | HB | 335 247* | HH |
| 335 187* | FG | 335 203* | AH4 | 335 218* | EOB | 335 233* | EHM | 335 248* | HB |
| 335 188* | HB | 335 204* | EHG | 335 219* | EOB | 335 234* | AL | 335 249* | HH |
| 335 189* | FK | 335 205* | KM | 335 220* | EOB | 335 235* | AL | 335 250* | HH |
| 335 190* | HB | 335 206* | KM | 335 221* | EHM | 335 236* | AL | 335 251* | HB |
| 335 191* | HBS | | | | | | | | |

## CLASS 344         D

**Built:** 1959-83 by LKM/LEW as classes 105 & 106. Rebuilt 1991 onwards by DR Chemnitz Works.
**Engine:** Johannisthal 12 KVD 18/21 SWV3 of 365 kW.
**Transmission:** Hydraulic.        **Maximum Speed:** 44 km/h.
**Weight:**        **Length:**

| | | | | | | | | | |
|---|---|---|---|---|---|---|---|---|---|
| 344 005 | DR | 344 063 | UEI | 344 163 | LH1 | 344 808 | WR | 344 915 | DC |
| 344 006 | WW | 344 066 | WR | 344 669 | BHF | 344 827 | DZW | 344 923 | BHF |
| 344 007 | UE | 344 071 | BSC | 344 681 | WP | 344 832 | UE | 344 940 | DC |
| 344 009 | DG | 344 078 | LHB | 344 682 | BWUR | 344 835 | DC | 344 952 | UE |
| 344 011 | WNT | 344 081 | US | 344 687 | LL1 | 344 836 | BSC | 344 955 | WP |
| 344 014 | DR | 344 084 | LHB | 344 688 | BSC | 344 855 | DC | 344 958 | WS |
| 344 015 | DG | 344 087 | DR | 344 710 | BSC | 344 856 | BHF | 344 961 | UM |
| 344 018 | LHB | 344 090 | WW | 344 720 | WP | 344 866 | LL1 | 344 964 | DZW |
| 344 019 | WW | 344 092 | BWUR | 344 736 | LH1 | 344 868 | LL2 | 344 966 | DZW |
| 344 025 | LMR | 344 093 | WS | 344 752 | LS | 344 886 | DC | 344 968 | LL2 |
| 344 032 | DG | 344 103 | WW | 344 764 | BSC | 344 900 | LH1 | 344 976 | UN |
| 344 034 | US | 344 106 | BHF | 344 767 | LS | 344 903 | LHB | 344 986 | DH |
| 344 045 | LL1 | 344 111 | BSC | 344 768 | DH | 344 905 | BHF | 344 987 | BHF |
| 344 054 | LL2 | 344 125 | LH1 | 344 773 | WS | 344 910 | LH1 | 344 990 | DR |
| 344 059 | LMR | 344 134 | LMR | 344 788 | WR | 344 913 | DZW | 344 996 | DH |
| 344 060 | LHB | 344 146 | LH1 | 344 804 | LMR | 344 914 | DZW | 344 999 | DH |
| 344 062 | BWUR | 344 160 | LH1 | | | | | | |

## CLASS 345         D

**Built:** 1965-83 by LEW. Locomotives numbered in the 345.9xx series were initially in industrial use, but later purchased by the DR.
**Engine:** Johannisthal 12 KVD 21 SVW of 478 kW at 1500 rpm.
**Transmission:** Hydraulic. Strömungsmachinen GS 12/5.2.
**Weight:** 60.00 tonnes.        **Length:** 10.880 m.
**Note:** Previously DR Class 105.

| | | | | | | | | | |
|---|---|---|---|---|---|---|---|---|---|
| 345 001 | BSE | 345 023 | DR | 345 044 | DR | 345 064 | UEI | 345 083 | LL2 |
| 345 002 | LM | 345 026 | DG | 345 046 | WS | 345 065 | LMR | 345 085 | LHB |
| 345 003 | BHW | 345 028 | BPKR | 345 047 | DR | 345 067 | BHW | 345 088 | DR |
| 345 004 | WNT | 345 029 | BHF | 345 048 | WR | 345 068 | BPKR | 345 089 | LHB |
| 345 008 | BHF | 345 030 | LMR | 345 049 | DR | 345 069 | DH | 345 091 | DR |
| 345 010 | DR | 345 031 | BHF | 345 050 | UM | 345 070 | BHW | 345 094 | DG |
| 345 012 | LMR | 345 033 | WR | 345 051 | BHF | 345 072 | UN | 345 095 | UN |
| 345 013 | WP | 345 035 | DR | 345 052 | WSR | 345 073 | LH1 | 345 097 | LH1 |
| 345 016 | BWUR | 345 037 | LLW | 345 053 | LH1 | 345 075 | BFG | 345 098 | BSE |
| 345 017 | BFG | 345 038 | WW | 345 055 | LHB | 345 076 | US | 345 099 | LL1 |
| 345 020 | LMR | 345 040 | DG | 345 056 | WP | 345 077 | BHW | 345 100 | LMR |
| 345 021 | LL2 | 345 041 | (Z) | 345 057 | WR | 345 080 | LL1 | 345 101 | LMR |
| 345 022 | US | 345 042 | UE | 345 061 | LMR | 345 082 | DR | 345 102 | BHF |

| | | | | | | | | |
|---|---|---|---|---|---|---|---|---|---|
| 345 104 | UG | 345 118 | BHF | 345 132 | LH1 | 345 148 | LHB | 345 159 | LH1 |
| 345 105 | DH | 345 119 | BSC | 345 133 | LH1 | 345 149 | LH1 | 345 161 | LMR |
| 345 107 | US | 345 121 | BHF | 345 135 | BSC | 345 150 | LHB | 345 162 | LMR |
| 345 108 | BSE | 345 122 | LL2 | 345 136 | DR | 345 151 | LH1 | 345 164 | LH1 |
| 345 109 | BSE | 345 123 | LHB | 345 137 | LM | 345 152 | LH1 | 345 165 | LH1 |
| 345 110 | DR | 345 124 | LHB | 345 138 | LHB | 345 153 | WR | 345 965 | BHW |
| 345 112 | WW | 345 127 | LH1 | 345 139 | UG | 345 154 | WS | 345 966 | BSC |
| 345 113 | BSC | 345 128 | DR | 345 143 | WW | 345 155 | WR | 345 970 | DH |
| 345 114 | WP | 345 129 | US | 345 144 | WS | 345 156 | WW | 345 975 | UG |
| 345 115 | LHB | 345 130 | UEI | 345 145 | LH1 | 345 157 | LH1 | 345 990 | LHB |
| 345 116 | LM | 345 131 | BWUR | 345 147 | LH1 | 345 158 | LH1 | 345 991 | LS |
| 345 117 | LW | | | | | | | | |

## CLASS 346.0      D

**Built:** 1959-64 by LKM. Locomotives 346 181-185 were initially in industrial use, but were later purchased by DR.
**Engine:** Johannisthal 12 KVD 21 SVW of 478 kW at 1500 rpm.
**Transmission:** Hydraulic. Strömungsmaschinen GS 12/5.1.
**Weight:** 55.00 tonnes.      **Maximum Speed:** 60 km/h.
**Length:** 10.880 m.      **Note:** Previously DR Class 106.

| | | | | | | | | | |
|---|---|---|---|---|---|---|---|---|---|
| 346 012 | (Z) | 346 070 | (Z) | 346 100 | UW | 346 181 | BWUR | 346 185 | UEI |
| 346 065 | (Z) | 346 082 | (Z) | 346 153 | (Z) | 346 183 | BFG | | |

## CLASS 346.2      D

**Built:** 1965-83 by LEW.
**Engine:** Johannisthal 12 KVD 21 SVW of 478 kW at 1500 rpm.
**Transmission:** Hydraulic. Strömungsmaschinen GS 12/5.2.
**Weight:** 60.00 tonnes.      **Maximum Speed:** 60 km/h.
**Length:** 10.880 m.      **Note:** Previously DR Class 106.

| | | | | | | | | | |
|---|---|---|---|---|---|---|---|---|---|
| 346 201 | BSE | 346 245 | (Z) | 346 302 | DZW | 346 350 | DH | 346 394 | LLW |
| 346 202 | BSC | 346 249 | (Z) | 346 303 | (Z) | 346 351 | DH | 346 395 | LLW |
| 346 204 | LLW | 346 250 | BWUR | 346 305 | LMR | 346 352 | (Z) | 346 396 | UE |
| 346 205 | LLW | 346 251 | BFG | 346 306 | BPKR | 346 353 | (Z) | 346 397 | BSE |
| 346 206 | UN | 346 255 | (Z) | 346 307 | BPKR | 346 354 | LW | 346 398 | BSE |
| 346 210 | (Z) | 346 257 | DH | 346 310 | UE | 346 355 | LL1 | 346 399 | WNT |
| 346 212 | WNT | 346 259 | LMR | 346 311 | LMR | 346 356 | (Z) | 346 400 | DZW |
| 346 213 | UE | 346 261 | DH | 346 312 | US | 346 357 | (Z) | 346 402 | WR |
| 346 215 | BWUR | 346 263 | UW | 346 314 | BWUR | 346 358 | (Z) | 346 403 | LLW |
| 346 216 | BHF | 346 267 | BWUR | 346 315 | LW | 346 359 | US | 346 404 | UEI |
| 346 217 | LW | 346 268 | BHF | 346 316 | LLW | 346 360 | BPKR | 346 405 | DH |
| 346 218 | LW | 346 269 | BWUR | 346 317 | UE | 346 361 | BSE | 346 406 | UEI |
| 346 219 | UW | 346 270 | (Z) | 346 318 | (Z) | 346 362 | (Z) | 346 407 | UEI |
| 346 220 | LW | 346 273 | DH | 346 319 | DH | 346 363 | BHW | 346 408 | BHW |
| 346 221 | DH | 346 274 | LL1 | 346 320 | LLW | 346 364 | DH | 346 411 | (Z) |
| 346 223 | LLW | 346 275 | WNT | 346 322 | WP | 346 366 | DH | 346 412 | (Z) |
| 346 224 | LW | 346 276 | DH | 346 323 | LL1 | 346 367 | WP | 346 413 | UE |
| 346 226 | LL1 | 346 277 | BWUR | 346 325 | LH1 | 346 369 | DC | 346 414 | UW |
| 346 227 | LW | 346 278 | BWUR | 346 326 | LL1 | 346 370 | BPKR | 346 415 | LLW |
| 346 229 | BSC | 346 282 | (Z) | 346 329 | (Z) | 346 371 | DH | 346 416 | LL2 |
| 346 231 | LW | 346 284 | BWUR | 346 330 | (Z) | 346 372 | DC | 346 417 | UE |
| 346 232 | LW | 346 285 | BWUR | 346 331 | DC | 346 373 | DH | 346 418 | UE |
| 346 233 | LW | 346 286 | UEI | 346 332 | LH1 | 346 376 | (Z) | 346 419 | LH1 |
| 346 234 | LW | 346 290 | DH | 346 333 | LLW | 346 377 | LW | 346 420 | WR |
| 346 235 | LLW | 346 291 | DZW | 346 334 | LH1 | 346 378 | LL1 | 346 421 | UW |
| 346 236 | LLW | 346 292 | LM | 346 338 | (Z) | 346 380 | LLW | 346 422 | LLW |
| 346 237 | LH1 | 346 293 | LMR | 346 339 | LW | 346 383 | UE | 346 428 | UEI |
| 346 240 | (Z) | 346 295 | LW | 346 340 | LL1 | 346 385 | BSE | 346 430 | DH |
| 346 241 | BPKR | 346 296 | LL1 | 346 342 | LL1 | 346 387 | BSE | 346 431 | LLW |
| 346 242 | (Z) | 346 297 | LW | 346 344 | (Z) | 346 388 | LLW | 346 432 | BSE |
| 346 243 | UE | 346 298 | BWUR | 346 345 | (Z) | 346 389 | LW | 346 434 | (Z) |
| 346 244 | UW | 346 299 | DZW | 346 347 | LL1 | 346 390 | LS | 346 436 | DH |

# GERMANY

▲ 346 831 at Bad Freienwalde (J. Hayes)

▼ 401 002 at Offenburg on 22.05.95 (J. Hayes)

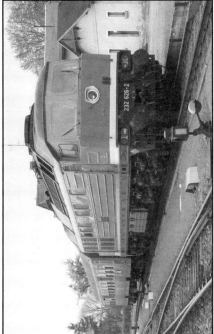

▲ 232 626 waits departure time from Schleiz with the 1220 to Plauen on 30.04.95 (M. Dunn)

▼ 298 305 at Magdeburg-Rothensee depot (Rail Europe)

| | | | | | | | | | |
|---|---|---|---|---|---|---|---|---|---|
| 346 437 | LH1 | 346 526 | DC | 346 598 | US | 346 673 | BFG | 346 744 | WP |
| 346 438 | LH1 | 346 528 | WS | 346 599 | BHF | 346 674 | WNT | 346 745 | LS |
| 346 439 | LLW | 346 529 | BFG | 346 600 | BSE | 346 675 | UG | 346 746 | (Z) |
| 346 440 | LL1 | 346 530 | LS | 346 601 | UE | 346 676 | WSR | 346 747 | LS |
| 346 441 | LLW | 346 531 | DH | 346 603 | BHW | 346 677 | BPKR | 346 748 | WP |
| 346 443 | (Z) | 346 532 | UG | 346 604 | DH | 346 678 | BHW | 346 749 | LM |
| 346 445 | DZW | 346 533 | UEI | 346 605 | DH | 346 679 | BPKR | 346 750 | DG |
| 346 446 | UE | 346 534 | BWUR | 346 606 | BHF | 346 680 | WP | 346 751 | WP |
| 346 447 | DH | 346 535 | UG | 346 607 | BPKR | 346 683 | BHW | 346 753 | BPKR |
| 346 448 | UM | 346 536 | (Z) | 346 609 | BHW | 346 684 | LMR | 346 754 | US |
| 346 450 | BSE | 346 538 | LS | 346 610 | DH | 346 685 | LH1 | 346 755 | BFG |
| 346 451 | DC | 346 539 | BWUR | 346 611 | BSE | 346 686 | BWUR | 346 756 | DZW |
| 346 452 | (Z) | 346 540 | BWUR | 346 612 | LS | 346 689 | WSR | 346 757 | BFG |
| 346 454 | (Z) | 346 541 | UG | 346 613 | LMR | 346 690 | LMR | 346 758 | UM |
| 346 457 | UEI | 346 543 | LL1 | 346 614 | WR | 346 691 | BFG | 346 759 | BHF |
| 346 458 | UE | 346 544 | UE | 346 615 | LS | 346 692 | WNT | 346 760 | BPKR |
| 346 459 | UN | 346 545 | LMR | 346 616 | BPKR | 346 693 | BPKR | 346 761 | WP |
| 346 462 | LH1 | 346 546 | (Z) | 346 617 | UE | 346 694 | BHW | 346 762 | DZW |
| 346 464 | LLW | 346 547 | BHF | 346 618 | WNT | 346 695 | LM | 346 763 | LS |
| 346 465 | US | 346 548 | BHF | 346 619 | WS | 346 696 | DG | 346 765 | BWUR |
| 346 466 | DC | 346 549 | WNT | 346 620 | BSE | 346 697 | WNT | 346 766 | BFG |
| 346 467 | WSR | 346 551 | BPKR | 346 621 | US | 346 698 | BFG | 346 769 | WSR |
| 346 468 | DC | 346 552 | DZW | 346 624 | LM | 346 699 | BPKR | 346 770 | BSC |
| 346 469 | (Z) | 346 553 | UN | 346 625 | BWUR | 346 700 | LM | 346 771 | UM |
| 346 470 | UN | 346 554 | (Z) | 346 626 | UEI | 346 701 | WSR | 346 772 | DZW |
| 346 473 | BSC | 346 555 | DG | 346 630 | UEI | 346 702 | WW | 346 774 | WNT |
| 346 474 | (Z) | 346 556 | LW | 346 631 | UEI | 346 703 | LS | 346 775 | DH |
| 346 476 | UN | 346 557 | BWUR | 346 632 | WSR | 346 704 | WNT | 346 776 | BHW |
| 346 477 | (Z) | 346 558 | US | 346 633 | WR | 346 705 | LM | 346 777 | LW |
| 346 478 | DZW | 346 559 | DZW | 346 636 | DG | 346 706 | BWUR | 346 778 | BHW |
| 346 479 | UEI | 346 560 | (Z) | 346 637 | DH | 346 707 | BPKR | 346 779 | DZW |
| 346 480 | US | 346 562 | WNT | 346 638 | LH1 | 346 708 | UEI | 346 780 | UEI |
| 346 482 | DH | 346 563 | (Z) | 346 639 | WSR | 346 709 | BPKR | 346 781 | WP |
| 346 483 | UEI | 346 564 | UN | 346 640 | BPKR | 346 711 | BPKR | 346 782 | LS |
| 346 485 | UM | 346 565 | DZW | 346 641 | BPKR | 346 712 | LMR | 346 783 | LM |
| 346 487 | UE | 346 566 | US | 346 642 | DG | 346 713 | WR | 346 784 | LMR |
| 346 489 | UN | 346 567 | (Z) | 346 643 | BPKR | 346 714 | BPKR | 346 785 | BHF |
| 346 492 | UEI | 346 568 | US | 346 644 | LLW | 346 715 | LMR | 346 786 | WS |
| 346 494 | DZW | 346 569 | LMR | 346 645 | LL1 | 346 717 | LMR | 346 789 | UN |
| 346 496 | LW | 346 570 | (Z) | 346 646 | WR | 346 718 | WR | 346 790 | DG |
| 346 497 | DZW | 346 571 | UG | 346 647 | WW | 346 719 | UM | 346 791 | LMR |
| 346 498 | WSR | 346 572 | LMR | 346 648 | US | 346 721 | WS | 346 792 | WW |
| 346 500 | LL2 | 346 573 | (Z) | 346 649 | LS | 346 722 | WS | 346 793 | LW |
| 346 502 | BPKR | 346 574 | BPKR | 346 650 | LMR | 346 723 | DH | 346 794 | LHB |
| 346 503 | DZW | 346 576 | BWUR | 346 651 | LMR | 346 724 | DG | 346 795 | BWUR |
| 346 504 | BPKR | 346 577 | BWUR | 346 652 | UN | 346 725 | WS | 346 797 | BFG |
| 346 505 | (Z) | 346 578 | BWUR | 346 653 | LMR | 346 726 | BSE | 346 798 | UEI |
| 346 506 | LL2 | 346 579 | DZW | 346 654 | LM | 346 727 | WNT | 346 799 | DG |
| 346 508 | DZW | 346 580 | DC | 346 655 | UEI | 346 728 | WS | 346 800 | DZW |
| 346 509 | (Z) | 346 581 | UE | 346 656 | BHF | 346 729 | LM | 346 801 | BFG |
| 346 511 | BHF | 346 582 | (Z) | 346 658 | DG | 346 730 | LM | 346 802 | DC |
| 346 512 | LM | 346 583 | (Z) | 346 659 | US | 346 731 | LM | 346 803 | (Z) |
| 346 513 | DC | 346 585 | (Z) | 346 661 | LMR | 346 732 | LHB | 346 805 | WNT |
| 346 514 | UG | 346 586 | BHW | 346 662 | DZW | 346 733 | BHW | 346 806 | LM |
| 346 516 | WR | 346 587 | BHW | 346 663 | BWUR | 346 734 | BHW | 346 807 | WS |
| 346 517 | BHW | 346 589 | BWUR | 346 664 | (Z) | 346 735 | WSR | 346 809 | BPKR |
| 346 518 | WSR | 346 590 | LMR | 346 665 | BHW | 346 737 | DZW | 346 810 | LHB |
| 346 519 | LL1 | 346 591 | BHF | 346 666 | LH1 | 346 738 | DH | 346 811 | WP |
| 346 520 | DH | 346 592 | DZW | 346 667 | (Z) | 346 739 | WR | 346 812 | WS |
| 346 521 | UE | 346 594 | (Z) | 346 668 | BHW | 346 740 | DH | 346 813 | DH |
| 346 522 | WW | 346 595 | BFG | 346 670 | DG | 346 741 | DC | 346 814 | UN |
| 346 523 | BHW | 346 596 | US | 346 671 | BPKR | 346 742 | LHB | 346 815 | WNT |
| 346 524 | BSE | 346 597 | DC | 346 672 | BHF | 346 743 | DR | 346 816 | WP |

# GERMANY

| | | | | | | | | | |
|---|---|---|---|---|---|---|---|---|---|
| 346 817 | LHB | 346 852 | DZW | 346 888 | LHB | 346 927 | LLW | 346 960 | (Z) |
| 346 818 | LL1 | 346 853 | WW | 346 889 | BFG | 346 928 | BPKR | 346 962 | UW |
| 346 819 | LLW | 346 854 | LS | 346 890 | BPKR | 346 929 | UEI | 346 963 | DH |
| 346 820 | LL1 | 346 857 | US | 346 891 | LL2 | 346 930 | LL1 | 346 965 | BPKR |
| 346 821 | UE | 346 858 | DG | 346 893 | DC | 346 931 | LH1 | 346 967 | WSR |
| 346 822 | DG | 346 859 | (Z) | 346 894 | (Z) | 346 932 | LHB | 346 969 | BFG |
| 346 823 | BPKR | 346 860 | WW | 346 895 | US | 346 933 | LW | 346 970 | WSR |
| 346 824 | DK | 346 861 | LL1 | 346 896 | BFG | 346 934 | BSE | 346 971 | WS |
| 346 825 | LH1 | 346 862 | DH | 346 897 | DC | 346 935 | DG | 346 972 | UM |
| 346 826 | LL1 | 346 863 | BPKR | 346 898 | (Z) | 346 936 | BHW | 346 973 | WS |
| 346 828 | BHW | 346 864 | WSR | 346 899 | UN | 346 937 | DH | 346 974 | BWUR |
| 346 829 | LHB | 346 865 | BSE | 346 901 | UM | 346 938 | DH | 346 977 | LM |
| 346 830 | LH1 | 346 867 | BHW | 346 902 | LHB | 346 939 | DC | 346 978 | LW |
| 346 831 | WP | 346 869 | DC | 346 904 | BPKR | 346 941 | LH1 | 346 979 | DZW |
| 346 833 | BHF | 346 870 | LMR | 346 906 | DZW | 346 942 | BSE | 346 980 | BHW |
| 346 834 | WNT | 346 872 | DH | 346 907 | LHB | 346 943 | UN | 346 981 | DR |
| 346 837 | LLW | 346 873 | WS | 346 908 | LS | 346 944 | LW | 346 982 | DR |
| 346 838 | LL2 | 346 874 | LH1 | 346 909 | LMR | 346 945 | BHW | 346 983 | WS |
| 346 839 | WNT | 346 875 | BSC | 346 911 | BPKR | 346 946 | BHW | 346 984 | WS |
| 346 840 | WNT | 346 876 | DG | 346 912 | BHF | 346 947 | WSR | 346 985 | DR |
| 346 841 | (Z) | 346 877 | LS | 346 916 | BSE | 346 948 | LS | 346 988 | UN |
| 346 843 | BHW | 346 878 | WSR | 346 917 | BFG | 346 949 | WP | 346 989 | BHW |
| 346 844 | UE | 346 879 | UEI | 346 918 | DH | 346 950 | LL2 | 346 991 | LL1 |
| 346 845 | WP | 346 880 | BPKR | 346 919 | UN | 346 951 | WR | 346 992 | DZW |
| 346 846 | LL1 | 346 881 | LM | 346 920 | LH1 | 346 953 | UEI | 346 993 | WS |
| 346 847 | WW | 346 882 | LM | 346 921 | BHF | 346 954 | LHB | 346 994 | LM |
| 346 848 | WR | 346 883 | DH | 346 922 | LMR | 346 956 | LL2 | 346 995 | DZW |
| 346 849 | BSE | 346 884 | BHF | 346 924 | DC | 346 957 | LH1 | 346 997 | (Z) |
| 346 850 | DH | 346 885 | BPKR | 346 925 | BHF | 346 959 | DC | 346 998 | (Z) |
| 346 851 | BPKR | 346 887 | DC | 346 926 | BSE | | | | |

## CLASS 347   D

Technical details as Class 346 except:
**Gauge:** 1524 mm.
**Note:** Sub shedded at Mukran and used for shunting the train ferry terminal there.

| | | | | | | | | | |
|---|---|---|---|---|---|---|---|---|---|
| 347 024 | (Z) | 347 079 | (Z) | 347 140 | WSR | 347 142 | (Z) | 347 975 | (Z) |
| 347 036 | WSR | 347 096 | (Z) | 347 141 | WSR | | | | |

## CLASS 360   B

**Built:** 1956-62 by Esslingen/Krupp/Krauss-Maffei/Jung/Gmeinder/Henschel/Deutz/MaK.
**Engine:** Maybach GTO 6/Deutz BA128M 16/Mercedes MB 12 V 493 AZ of 480/492 kW at 1400 rpm.
**Transmission:** Hydraulic. Voith L37zUb/Voith L27zUb/Voith L217.
**Weight:** 48.00-50.00 tonnes.     **Maximum Speed:** 60 km/h.
**Length:** 10.450 m.

| | | | | | | | | |
|---|---|---|---|---|---|---|---|---|
| 360 002 | AFL | 360 032 | FF2 | 360 114 | NHO | 360 137 | MH1 | 360 156 | RK |
| 360 004 | AH4 | 360 033 | FF2 | 360 115 | NN1 | 360 138 | AH4 | 360 157 | NN1 |
| 360 006 | EDO | 360 034 | FF2 | 360 116 | RHL | 360 140 | HE | 360 158 | NN1 |
| 360 007 | AFL | 360 035 | FF2 | 360 117 | MH1 | 360 141 | RHL | 360 159 | NWH |
| 360 008 | AH4 | 360 046 | HH | 360 118 | NN1 | 360 142 | MH1 | 360 160 | AH4 |
| 360 009 | EOB | 360 047 | ESIE | 360 119 | MH1 | 360 143 | MH1 | 360 161 | NWH |
| 360 010 | TU | 360 048 | FK | 360 123 | EOB | 360 146 | NWH | 360 162 | NWH |
| 360 012 | EDO | 360 102 | ESIE | 360 125 | NRH | 360 147 | NRH | 360 163 | HE |
| 360 013 | AFL | 360 104 | MH1 | 360 126 | NRH | 360 148 | NN1 | 360 165 | NWH |
| 360 014 | EDO | 360 105 | HE | 360 127 | NN1 | 360 149 | KK | 360 166 | RM |
| 360 015 | EDO | 360 107 | EHM | 360 130 | MH1 | 360 150 | NN1 | 360 167 | EDO |
| 360 017 | EOB | 360 109 | AH4 | 360 131 | MH1 | 360 151 | NN1 | 360 169 | AL |
| 360 019 | FK | 360 110 | AH4 | 360 132 | MH1 | 360 152 | RHL | 360 171 | RM |
| 360 022 | AH4 | 360 111 | KK | 360 133 | MH1 | 360 153 | RO | 360 172 | MH1 |
| 360 023 | AFL | 360 112 | AH4 | 360 134 | MH1 | 360 154 | SKL | 360 173 | AH4 |
| 360 027 | FF2 | 360 113 | NN1 | 360 135 | AH4 | 360 155 | EDO | 360 174 | AH4 |

| Number | Code | Number | Code | Number | Code | Number | Code | Number | Code |
|---|---|---|---|---|---|---|---|---|---|
| 360 176 | STR | 360 280 | AL | 360 362 | AH4 | 360 522 | EDO | 360 754 | SSH |
| 360 177 | HG | 360 281 | EOB | 360 363 | AH4 | 360 525 | EHM | 360 755 | SSH |
| 360 178 | NWH | 360 282 | EOB | 360 364 | AL | 360 526 | RSI | 360 756 | SSH |
| 360 180 | EDO | 360 283 | AFL | 360 365 | AL | 360 533 | FG | 360 757 | SKL |
| 360 201 | TU | 360 284 | AFL | 360 366 | TK | 360 534 | FG | 360 758 | FF2 |
| 360 202 | TK | 360 285 | EOB | 360 367 | AH4 | 360 537 | FF2 | 360 759 | FK |
| 360 204 | RSI | 360 286 | AH4 | 360 368 | AFL | 360 538 | KK | 360 760 | EHM |
| 360 205 | MA | 360 288 | AL | 360 370 | EOB | 360 540 | RO | 360 761 | FK |
| 360 207 | MA | 360 289 | AL | 360 371 | AH4 | 360 541 | RM | 360 763 | FK |
| 360 208 | MA | 360 290 | FG | 360 372 | RSI | 360 542 | RK | 360 765 | FF2 |
| 360 209 | NRH | 360 292 | FG | 360 373 | AH4 | 360 543 | RO | 360 768 | TK |
| 360 210 | NN1 | 360 293 | EOB | 360 374 | AH4 | 360 546 | RO | 360 770 | FD |
| 360 211 | NWH | 360 294 | FG | 360 376 | NWH | 360 547 | EOB | 360 772 | FF2 |
| 360 213 | MA | 360 296 | FG | 360 377 | EOB | 360 550 | FD | 360 773 | FD |
| 360 214 | NN1 | 360 297 | RM | 360 378 | SKL | 360 551 | KK | 360 774 | FF2 |
| 360 215 | AH4 | 360 298 | HE | 360 379 | EOB | 360 555 | RK | 360 775 | FF2 |
| 360 216 | EOB | 360 300 | AH4 | 360 382 | FF2 | 360 563 | FD | 360 777 | TK |
| 360 217 | EOB | 360 303 | HBS | 360 383 | FD | 360 564 | KK | 360 778 | TK |
| 360 219 | NRH | 360 304 | EDO | 360 385 | AH4 | 360 565 | KK | 360 781 | RM |
| 360 220 | NRH | 360 307 | MH1 | 360 386 | MH1 | 360 567 | KK | 360 782 | RM |
| 360 221 | NRH | 360 308 | HO | 360 387 | STR | 360 568 | EHM | 360 783 | RM |
| 360 222 | AH4 | 360 309 | FF2 | 360 388 | SKL | 360 570 | EOB | 360 784 | FF2 |
| 360 223 | NRH | 360 311 | HH | 360 389 | EOB | 360 571 | KK | 360 785 | RM |
| 360 225 | NHO | 360 312 | HH | 360 390 | SKL | 360 572 | EHG | 360 787 | TK |
| 360 227 | EOB | 360 314 | HH | 360 391 | HBH | 360 573 | EHM | 360 788 | TK |
| 360 229 | AH4 | 360 315 | HH | 360 392 | HBH | 360 576 | MH1 | 360 789 | TK |
| 360 231 | NN1 | 360 316 | HH | 360 401 | EHM | 360 577 | EHM | 360 790 | TK |
| 360 232 | RK | 360 317 | HBS | 360 402 | KK | 360 579 | NN1 | 360 791 | KK |
| 360 233 | NHO | 360 318 | EDO | 360 404 | KK | 360 580 | HE | 360 792 | RK |
| 360 234 | NN1 | 360 319 | AL | 360 406 | AH4 | 360 582 | HH | 360 793 | RM |
| 360 235 | NRH | 360 320 | AL | 360 407 | AH4 | 360 583 | EHG | 360 795 | KK |
| 360 236 | NN1 | 360 321 | AL | 360 408 | KK | 360 587 | EOB | 360 798 | TK |
| 360 237 | NHO | 360 322 | AL | 360 409 | KK | 360 588 | TK | 360 800 | HH |
| 360 239 | NN1 | 360 323 | AH4 | 360 410 | KK | 360 589 | KK | 360 802 | FF2 |
| 360 240 | AFL | 360 324 | EHM | 360 411 | AH4 | 360 591 | EHG | 360 803 | RHL |
| 360 241 | NHO | 360 325 | EHM | 360 412 | AH4 | 360 592 | EHG | 360 804 | KK |
| 360 244 | NHO | 360 326 | MH1 | 360 413 | NHO | 360 593 | KK | 360 806 | MMF |
| 360 245 | NN1 | 360 327 | MH1 | 360 414 | EDO | 360 594 | EHG | 360 807 | SSH |
| 360 246 | NHO | 360 328 | NN1 | 360 415 | EHM | 360 595 | FD | 360 842 | MH1 |
| 360 247 | NN1 | 360 329 | TK | 360 416 | MH1 | 360 596 | KK | 360 843 | NHO |
| 360 248 | NHO | 360 330 | EDO | 360 417 | MH1 | 360 597 | KK | 360 844 | MH1 |
| 360 249 | MH1 | 360 332 | TK | 360 419 | HH | 360 599 | HO | 360 846 | MH1 |
| 360 251 | AH4 | 360 333 | TK | 360 420 | EHG | 360 600 | EHM | 360 849 | MMF |
| 360 252 | AH4 | 360 334 | EOB | 360 421 | FF2 | 360 603 | NWH | 360 853 | MA |
| 360 253 | AH4 | 360 335 | TK | 360 422 | FG | 360 604 | AH4 | 360 855 | NHO |
| 360 255 | AH4 | 360 336 | TK | 360 423 | TU | 360 606 | TK | 360 856 | MH1 |
| 360 256 | AH4 | 360 338 | TK | 360 427 | AH4 | 360 607 | AH4 | 360 858 | MH1 |
| 360 257 | AH4 | 360 339 | TU | 360 428 | EHM | 360 608 | KK | 360 859 | MH1 |
| 360 259 | EHM | 360 340 | TU | 360 430 | RSI | 360 610 | KK | 360 860 | MH1 |
| 360 260 | AH4 | 360 341 | TU | 360 431 | NN1 | 360 612 | KK | 360 861 | MH1 |
| 360 261 | AH4 | 360 342 | TU | 360 449 | SSH | 360 613 | EDO | 360 864 | NHO |
| 360 262 | AH4 | 360 343 | TU | 360 501 | SSH | 360 614 | ESIE | 360 865 | NN1 |
| 360 264 | AL | 360 345 | RK | 360 502 | HO | 360 615 | ESIE | 360 866 | MH1 |
| 360 266 | ESIE | 360 346 | KK | 360 504 | RHL | 360 616 | HB | 360 868 | EHM |
| 360 267 | AH4 | 360 347 | AFL | 360 505 | RM | 360 617 | EDO | 360 870 | NHO |
| 360 268 | AH4 | 360 351 | FD | 360 506 | RM | 360 618 | EHM | 360 871 | HBS |
| 360 269 | AH4 | 360 352 | RK | 360 507 | MIH | 360 741 | NRH | 360 874 | NWH |
| 360 270 | AH4 | 360 353 | RK | 360 508 | MMF | 360 742 | NRH | 360 877 | MH1 |
| 360 271 | AH4 | 360 356 | EOB | 360 509 | AH4 | 360 743 | NRH | 360 878 | MH1 |
| 360 272 | AH4 | 360 357 | AL | 360 510 | FG | 360 744 | NRH | 360 883 | MH1 |
| 360 274 | EOB | 360 360 | AFL | 360 514 | TU | 360 746 | KK | 360 885 | NN1 |
| 360 277 | EOB | 360 361 | AH4 | 360 517 | AH4 | 360 750 | FK | 360 886 | NN1 |
| 360 278 | AL | | | 360 518 | HH | 360 753 | FK | 360 889 | NN1 |

# GERMANY

| | | | | | | | | | |
|---|---|---|---|---|---|---|---|---|---|
| 360 891 | NN1 | 360 904 | HO | 360 918 | NWH | 360 925 | FF2 | 360 934 | KK |
| 360 893 | FF2 | 360 908 | AH4 | 360 919 | NWH | 360 926 | FF2 | 360 935 | RM |
| 360 894 | FK | 360 911 | HO | 360 920 | NWH | 360 927 | FF2 | 360 936 | RM |
| 360 895 | RK | 360 913 | RM | 360 921 | NHO | 360 928 | EHG | 360 938 | AL |
| 360 896 | FD | 360 914 | EHM | 360 922 | MH1 | 360 929 | EHG | 360 939 | RM |
| 360 900 | HBS | 360 916 | HG | 360 923 | FD | 360 932 | KK | 360 941 | RM |
| 360 901 | HO | 360 917 | NWH | 360 924 | FF2 | 360 933 | KK | 360 942 | KK |
| 360 903 | AH4 | | | | | | | | |

## CLASS 361 B

**Built:** 1958-64 by Esslingen/Krupp/Krauss-Maffei/Jung/Gmeinder/Henschel/Deutz/MaK.
**Engine:** Maybach GTO 6/Deutz BA128M 16/Mercedes MB 12 V 493 AZ of 480/492 kW at 1400 rpm.
**Transmission:** Hydraulic. Voith L37zUb/Voith L27zUb/Voith L217.
**Weight:** 54.00 tonnes.  **Maximum Speed:** 60 km/h.
**Length:** 10.450 m.

| | | | | | | | | | |
|---|---|---|---|---|---|---|---|---|---|
| 361 036 | FF2 | 361 187 | HBS | 361 197 | AH4 | 361 232 | HH | 361 661 | EHM |
| 361 126 | RO | 361 188 | EHG | 361 204 | RF | 361 233 | RM | 361 663 | EDO |
| 361 149 | TK | 361 191 | HO | 361 207 | HB | 361 234 | HH | 361 692 | HBS |
| 361 165 | HH | 361 192 | HO | 361 221 | TK | 361 235 | HH | 361 808 | TK |
| 361 186 | EHG | 361 193 | AH4 | 361 222 | TK | 361 658 | HBH | 361 835 | EOB |

## CLASS 364 B

Technical details as for Class 360. Fitted with radio operated remote control.

| | | | | | | | | | |
|---|---|---|---|---|---|---|---|---|---|
| 364 400 | HE | 364 558 | MMF | 364 767 | HBS | 364 850 | MIH | 364 882 | MH1 |
| 364 403 | FG | 364 559 | HH | 364 769 | HBS | 364 851 | FF2 | 364 884 | NN1 |
| 364 447 | KK | 364 560 | RSI | 364 771 | FF2 | 364 852 | HH | 364 888 | NN1 |
| 364 448 | KK | 364 562 | AH4 | 364 776 | FG | 364 854 | MH1 | 364 890 | EDO |
| 364 450 | RHL | 364 566 | KK | 364 779 | FF2 | 364 857 | NN1 | 364 892 | HE |
| 364 511 | EHM | 364 569 | FF2 | 364 780 | TK | 364 862 | MIH | 364 902 | HH |
| 364 520 | KK | 364 574 | HH | 364 786 | TK | 364 863 | AH4 | 364 906 | HO |
| 364 523 | HO | 364 578 | HO | 364 794 | RSI | 364 867 | NN1 | 364 909 | HB |
| 364 524 | HB | 364 586 | FF2 | 364 796 | RM | 364 869 | FK | 364 910 | HO |
| 364 531 | FG | 364 598 | HH | 364 797 | TK | 364 872 | FK | 364 912 | HO |
| 364 535 | FF2 | 364 605 | HB | 364 799 | HH | 364 873 | MMF | 364 915 | HH |
| 364 536 | RSI | 364 611 | KK | 364 801 | EHG | 364 875 | MMF | 364 930 | TK |
| 364 544 | FF2 | 364 748 | HB | 364 805 | NN1 | 364 876 | MH1 | 364 937 | FK |
| 364 545 | HE | 364 751 | FK | 364 845 | MH1 | 364 879 | MH1 | 364 940 | RHL |
| 364 556 | HH | 364 762 | FK | 364 847 | MH1 | 364 881 | NRH | 364 943 | FF2 |
| 364 557 | RSI | 364 766 | HE | 364 848 | MMF | | | | |

## CLASS 365 B

Technical details as for Class 361. Fitted with radio operated remote control.

| | | | | | | | | | |
|---|---|---|---|---|---|---|---|---|---|
| 365 042 | RM | 365 114 | NRH | 365 132 | MH1 | 365 150 | SSH | 365 169 | TK |
| 365 043 | FF2 | 365 115 | TK | 365 133 | MH1 | 365 151 | FF2 | 365 170 | FG |
| 365 044 | RM | 365 116 | MH1 | 365 134 | AL | 365 152 | FK | 365 171 | NWH |
| 365 045 | RM | 365 117 | SSH | 365 135 | MMF | 365 153 | FG | 365 172 | SSH |
| 365 101 | RM | 365 118 | NWH | 365 136 | TK | 365 154 | FK | 365 173 | SSH |
| 365 102 | RK | 365 119 | NRH | 365 137 | KK | 365 155 | FK | 365 174 | HBS |
| 365 103 | RK | 365 120 | MIH | 365 138 | FF2 | 365 156 | AH4 | 365 175 | SSH |
| 365 104 | RK | 365 121 | FF2 | 365 139 | TK | 365 158 | AH4 | 365 176 | KK |
| 365 105 | FF2 | 365 122 | FF2 | 365 140 | STR | 365 159 | AH4 | 365 177 | FG |
| 365 106 | FF2 | 365 123 | MH1 | 365 141 | SKL | 365 160 | AH4 | 365 178 | AH4 |
| 365 107 | NWH | 365 124 | TK | 365 142 | FK | 365 161 | TK | 365 179 | MKP |
| 365 108 | TK | 365 125 | NRH | 365 143 | TU | 365 162 | EOB | 365 180 | MH1 |
| 365 109 | HG | 365 127 | EOB | 365 144 | RM | 365 163 | HE | 365 181 | AL |
| 365 110 | NWH | 365 128 | EDO | 365 145 | RM | 365 164 | TK | 365 182 | MMF |
| 365 111 | HG | 365 129 | NRH | 365 146 | SSH | 365 166 | FG | 365 183 | AL |
| 365 112 | NWH | 365 130 | EOB | 365 147 | SSH | 365 167 | KK | 365 184 | AL |
| 365 113 | FF2 | 365 131 | EOB | 365 148 | SSH | 365 168 | NN1 | 365 185 | HG |

| | | | | | | | | | |
|---|---|---|---|---|---|---|---|---|---|
| 365 189 | RHL | 365 237 | MIH | 365 646 | RO | 365 699 | RM | 365 737 | TK |
| 365 190 | HG | 365 238 | AH4 | 365 647 | TK | 365 701 | SSH | 365 738 | TK |
| 365 194 | RO | 365 239 | AH4 | 365 649 | HB | 365 702 | AH4 | 365 739 | EOB |
| 365 195 | RSI | 365 240 | SKL | 365 650 | HG | 365 703 | SSH | 365 740 | EHM |
| 365 196 | TK | 365 241 | MIH | 365 651 | TU | 365 704 | MKP | 365 809 | EHG |
| 365 198 | NHO | 365 424 | SKL | 365 652 | HB | 365 706 | MKP | 365 810 | TU |
| 365 199 | NRH | 365 425 | HH | 365 653 | TU | 365 707 | MKP | 365 811 | TK |
| 365 200 | FF2 | 365 426 | HO | 365 654 | NN1 | 365 708 | MIH | 365 812 | EOB |
| 365 201 | RK | 365 434 | HB | 365 655 | HB | 365 709 | AL | 365 813 | AH4 |
| 365 202 | RM | 365 435 | HB | 365 656 | HBS | 365 710 | FF2 | 365 814 | MIH |
| 365 203 | FK | 365 436 | HB | 365 660 | KK | 365 711 | MMF | 365 815 | EDO |
| 365 205 | NN1 | 365 437 | KK | 365 662 | RM | 365 712 | HH | 365 816 | RM |
| 365 206 | HBS | 365 438 | HH | 365 664 | RM | 365 713 | RM | 365 817 | TK |
| 365 208 | NHO | 365 439 | MIH | 365 665 | RSI | 365 714 | AL | 365 818 | AH4 |
| 365 209 | NHO | 365 440 | EHM | 365 666 | RK | 365 715 | AL | 365 819 | TU |
| 365 210 | KK | 365 441 | AL | 365 668 | KK | 365 716 | TU | 365 820 | EDO |
| 365 211 | KK | 365 444 | FF2 | 365 669 | RM | 365 717 | FF2 | 365 821 | TU |
| 365 212 | FF2 | 365 445 | RM | 365 674 | FK | 365 718 | MH1 | 365 822 | FG |
| 365 213 | FK | 365 446 | NRH | 365 675 | FF2 | 365 719 | KK | 365 823 | FG |
| 365 214 | FK | 365 621 | EDO | 365 676 | AL | 365 720 | FF2 | 365 824 | TK |
| 365 215 | EOB | 365 622 | ESIE | 365 677 | AL | 365 721 | SKL | 365 825 | STR |
| 365 216 | NHO | 365 623 | NHO | 365 678 | MMF | 365 722 | STR | 365 826 | FK |
| 365 217 | FK | 365 624 | NHO | 365 679 | FF2 | 365 723 | TK | 365 827 | TU |
| 365 218 | TK | 365 625 | FK | 365 680 | SSH | 365 724 | NRH | 365 828 | FL |
| 365 219 | TK | 365 626 | NN1 | 365 681 | TK | 365 726 | EHG | 365 829 | TK |
| 365 220 | TU | 365 627 | NHO | 365 682 | EHM | 365 727 | STR | 365 830 | EDO |
| 365 223 | KK | 365 628 | NHO | 365 683 | TU | 365 728 | SKL | 365 831 | HBS |
| 365 224 | KK | 365 631 | NWH | 365 684 | EHM | 365 729 | EHM | 365 832 | TK |
| 365 225 | EOB | 365 632 | NWH | 365 685 | SSH | 365 730 | RK | 365 833 | EOB |
| 365 226 | FG | 365 633 | NN1 | 365 686 | SKL | 365 731 | RK | 365 834 | EHG |
| 365 227 | FK | 365 634 | MMF | 365 688 | SSH | 365 732 | EOB | 365 836 | TK |
| 365 228 | FF2 | 365 635 | NN1 | 365 689 | TU | 365 733 | EOB | 365 837 | SKL |
| 365 229 | FG | 365 636 | NRH | 365 691 | EHM | 365 734 | ESIE | 365 838 | HB |
| 365 230 | NRH | 365 638 | HH | 365 695 | TK | 365 735 | ESIE | 365 839 | HG |
| 365 231 | NRH | 365 641 | RSI | 365 696 | EHM | 365 736 | EHG | 365 840 | TK |
| 365 236 | HH | 365 642 | HH | 365 698 | EDO | | | | |

# BATTERY ELECTRIC SHUNTING LOCOMOTIVES

## CLASS 381          B

**Built:** 1938 by Windhoff/Siemens.
**Battery:** 6 GiS 400 of 360 A/h.      **Length:** 6.450 m.
**Weight:** 17.00 tonnes.      **Maximum Speed:** 25 km/h.

381 018 KOPLX      381 020   FF2

## CLASS 382          B

**Built:** 1954 by Gmeinder/Garbe-Lahmeyer.
**Battery:** 6 GiS 400 of 360 A/h.      **Length:** 6.432 m.
**Weight:** 21.00 tonnes.      **Maximum Speed:** 30 km/h.

382 001 AOP

## CLASS 383          B

**Built:** 19?? by LEW. Acquired from Leune Werke AG in 1993.
**Battery:**      **Length:**
**Weight:** 12.00 tonnes.      **Maximum Speed:** 6 km/h.

383 001 KOPLX

# 1000 MM GAUGE DIESEL SHUNTING LOCOMOTIVES

## 399 101-103             C

**Built:** 1957 (* 1952) by Gmeinder.
**Engine:** Deutz A 6 M 617 of 94 kW at 1500 rpm.
**Transmission:** Hydraulic. Voith L22.     **Length:** 5.556 m.
**Weight:** 16.50 tonnes.     **Maximum Speed:** 20 km/h.

399 101* HWG      399 102  HWG      399 103  HWG

## 399 104             B

**Built:** 1952 by Deutz. Acquired from Inselbahn Juist (Heinrich) in 1973.
**Engine:** Deutz A 6 M 517 of 87 kW at 1500 rpm.
**Transmission:** Hydraulic. Voith L22.     **Length:** 6.090 m.
**Weight:** 11.50 tonnes.     **Maximum Speed:** 18 km/h.

399 104 HWG

## 399 105/6             C

**Built:** 1989 by 23 August. Acquired from Rohhütten GmbH, Helbra, in 1992.
**Engine:** MAN D2156 HMN 8 of 132 kW.
**Transmission:** Hydraulic.     **Length:**
**Weight:** 16.50 tonnes.     **Maximum Speed:** 28 km/h.

399 105 HWG      399 106  HWG

# 750 mm GAUGE DIESEL SHUNTING LOCOMOTIVES

## 399 703             C

**Built:** 1944 by Gmeinder.
**Engine:** Deutz A 6 M 517 of 77 kW at 1250 rpm.
**Transmission:** Hydraulic. Voith L22.     **Length:** 5.356 m.
**Weight:** 16.50 tonnes.     **Maximum Speed:** 20 km/h.

399 703 (Z)

# ELECTRIC MULTIPLE UNITS

## CLASS 401             ICE Power Car

**Built:** 1989-93 by Krupp/Krauss-Maffei/Henschel.
**Electrical Equipment:** Siemens/ABB/AEG.     **Weight:** 75.50 (* 78.00) tonnes.
**Continuous Rating:** 4800 kW.     **Length:** 20.560 m.
**Wheel Arrangement:** Bo-Bo.     **Maximum Speed:** 280 km/h.

| | | | | | | | | | |
|---|---|---|---|---|---|---|---|---|---|
| 401 001* | AH1 | 401 015* | AH1 | 401 059 | AH1 | 401 073 | AH1 | 401 087 | AH1 |
| 401 002* | AH1 | 401 016* | AH1 | 401 060 | AH1 | 401 074 | AH1 | 401 088 | AH1 |
| 401 003* | AH1 | 401 017* | AH1 | 401 061 | AH1 | 401 075 | AH1 | 401 089 | AH1 |
| 401 004* | AH1 | 401 018* | AH1 | 401 062 | AH1 | 401 076 | AH1 | 401 090 | AH1 |
| 401 005* | AH1 | 401 019* | AH1 | 401 063 | AH1 | 401 077 | AH1 | 401 501* | AH1 |
| 401 006* | AH1 | 401 020* | AH1 | 401 064 | AH1 | 401 078 | AH1 | 401 502* | AH1 |
| 401 007* | AH1 | 401 051 | AH1 | 401 065 | AH1 | 401 079 | AH1 | 401 503* | AH1 |
| 401 008* | AH1 | 401 052 | AH1 | 401 066 | AH1 | 401 080 | AH1 | 401 504* | AH1 |
| 401 009* | AH1 | 401 053 | AH1 | 401 067 | AH1 | 401 081 | AH1 | 401 505* | AH1 |
| 401 010* | AH1 | 401 054 | AH1 | 401 068 | AH1 | 401 082 | AH1 | 401 506* | AH1 |
| 401 011* | AH1 | 401 055 | AH1 | 401 069 | AH1 | 401 083 | AH1 | 401 507* | AH1 |
| 401 012* | AH1 | 401 056 | AH1 | 401 070 | AH1 | 401 084 | AH1 | 401 508* | AH1 |
| 401 013* | AH1 | 401 057 | AH1 | 401 071 | AH1 | 401 085 | AH1 | 401 509* | AH1 |
| 401 014* | AH1 | 401 058 | AH1 | 401 072 | AH1 | 401 086 | AH1 | 401 510* | AH1 |

| | | | | | | | | | |
|---|---|---|---|---|---|---|---|---|---|
| 401 511* | AH1 | 401 551 | AH1 | 401 561 | AH1 | 401 571 | AH1 | 401 581 | AH1 |
| 401 512* | AH1 | 401 552 | AH1 | 401 562 | AH1 | 401 572 | AH1 | 401 582 | AH1 |
| 401 513* | AH1 | 401 553 | AH1 | 401 563 | AH1 | 401 573 | AH1 | 401 583 | AH1 |
| 401 514* | AH1 | 401 554 | AH1 | 401 564 | AH1 | 401 574 | AH1 | 401 584 | AH1 |
| 401 515* | AH1 | 401 555 | AH1 | 401 565 | AH1 | 401 575 | AH1 | 401 585 | AH1 |
| 401 516* | AH1 | 401 556 | AH1 | 401 566 | AH1 | 401 576 | AH1 | 401 586 | AH1 |
| 401 517* | AH1 | 401 557 | AH1 | 401 567 | AH1 | 401 577 | AH1 | 401 587 | AH1 |
| 401 518* | AH1 | 401 558 | AH1 | 401 568 | AH1 | 401 578 | AH1 | 401 588 | AH1 |
| 401 519* | AH1 | 401 559 | AH1 | 401 569 | AH1 | 401 579 | AH1 | 401 589 | AH1 |
| 401 520* | AH1 | 401 560 | AH1 | 401 570 | AH1 | 401 580 | AH1 | 401 590 | AH1 |

## CLASS 801.0 — ICE Trailer First

**Built:** 1990-92 by Duewag.
**Wheel Arrangement:** 2-2.
**Weight:** 52.80 tonnes.
**Length:** 26.400 m.
**Seats:** 48F.
**Maximum Speed:** 280 km/h.

| | | | | | | | | | |
|---|---|---|---|---|---|---|---|---|---|
| 801 001 | AH1 | 801 021 | AH1 | 801 041 | AH1 | 801 061 | AH1 | 801 080 | AH1 |
| 801 002 | AH1 | 801 022 | AH1 | 801 042 | AH1 | 801 062 | AH1 | 801 081 | AH1 |
| 801 003 | AH1 | 801 023 | AH1 | 801 043 | AH1 | 801 063 | AH1 | 801 082 | AH1 |
| 801 004 | AH1 | 801 024 | AH1 | 801 044 | AH1 | 801 064 | AH1 | 801 083 | AH1 |
| 801 005 | AH1 | 801 025 | AH1 | 801 045 | AH1 | 801 065 | AH1 | 801 084 | AH1 |
| 801 006 | AH1 | 801 026 | AH1 | 801 046 | AH1 | 801 066 | AH1 | 801 085 | AH1 |
| 801 007 | AH1 | 801 027 | AH1 | 801 047 | AH1 | 801 067 | AH1 | 801 086 | AH1 |
| 801 008 | AH1 | 801 028 | AH1 | 801 048 | AH1 | 801 068 | AH1 | 801 087 | AH1 |
| 801 009 | AH1 | 801 029 | AH1 | 801 049 | AH1 | 801 069 | AH1 | 801 088 | AH1 |
| 801 010 | AH1 | 801 030 | AH1 | 801 050 | AH1 | 801 070 | AH1 | 801 089 | AH1 |
| 801 011 | AH1 | 801 031 | AH1 | 801 051 | AH1 | 801 071 | AH1 | 801 090 | AH1 |
| 801 012 | AH1 | 801 032 | AH1 | 801 052 | AH1 | 801 072 | AH1 | 801 091 | AH1 |
| 801 013 | AH1 | 801 033 | AH1 | 801 053 | AH1 | 801 073 | AH1 | 801 092 | AH1 |
| 801 014 | AH1 | 801 034 | AH1 | 801 054 | AH1 | 801 074 | AH1 | 801 093 | AH1 |
| 801 015 | AH1 | 801 035 | AH1 | 801 055 | AH1 | 801 075 | AH1 | 801 094 | AH1 |
| 801 016 | AH1 | 801 036 | AH1 | 801 056 | AH1 | 801 076 | AH1 | 801 095 | AH1 |
| 801 017 | AH1 | 801 037 | AH1 | 801 057 | AH1 | 801 077 | AH1 | 801 096 | AH1 |
| 801 018 | AH1 | 801 038 | AH1 | 801 058 | AH1 | 801 078 | AH1 | 801 097 | AH1 |
| 801 019 | AH1 | 801 039 | AH1 | 801 059 | AH1 | 801 079 | AH1 | 801 098 | AH1 |
| 801 020 | AH1 | 801 040 | AH1 | 801 060 | AH1 | | | | |

## CLASS 801.4 — ICE Trailer First

**Built:** 1990-92 by Waggon Union.
**Wheel Arrangement:** 2-2.
**Weight:** 52.80 tonnes.
Length: 26.400 m.
**Seats:** 48F.
**Maximum Speed:** 280 km/h.

| | | | | | | | | | |
|---|---|---|---|---|---|---|---|---|---|
| 801 401 | AH1 | 801 409 | AH1 | 801 417 | AH1 | 801 425 | AH1 | 801 433 | AH1 |
| 801 402 | AH1 | 801 410 | AH1 | 801 418 | AH1 | 801 426 | AH1 | 801 434 | AH1 |
| 801 403 | AH1 | 801 411 | AH1 | 801 419 | AH1 | 801 427 | AH1 | 801 435 | AH1 |
| 801 404 | AH1 | 801 412 | AH1 | 801 420 | AH1 | 801 428 | AH1 | 801 436 | AH1 |
| 801 405 | AH1 | 801 413 | AH1 | 801 421 | AH1 | 801 429 | AH1 | 801 437 | AH1 |
| 801 406 | AH1 | 801 414 | AH1 | 801 422 | AH1 | 801 430 | AH1 | 801 438 | AH1 |
| 801 407 | AH1 | 801 415 | AH1 | 801 423 | AH1 | 801 431 | AH1 | 801 439 | AH1 |
| 801 408 | AH1 | 801 416 | AH1 | 801 424 | AH1 | 801 432 | AH1 | 801 440 | AH1 |

## CLASS 801.8 — ICE Trailer First

**Built:** 1990-92 by Waggon Union.
**Wheel Arrangement:** 2-2.
**Weight:** 52.80 tonnes.
**Length:** 26.400 m.
**Seats:** 48F.
**Maximum Speed:** 280 km/h.

| | | | | | | | | | |
|---|---|---|---|---|---|---|---|---|---|
| 801 801 | AH1 | 801 805 | AH1 | 801 809 | AH1 | 801 813 | AH1 | 801 817 | AH1 |
| 801 802 | AH1 | 801 806 | AH1 | 801 810 | AH1 | 801 814 | AH1 | 801 818 | AH1 |
| 801 803 | AH1 | 801 807 | AH1 | 801 811 | AH1 | 801 815 | AH1 | 801 819 | AH1 |
| 801 804 | AH1 | 801 808 | AH1 | 801 812 | AH1 | 801 816 | AH1 | 801 820 | AH1 |

# GERMANY

| | | | | | | | | | |
|---|---|---|---|---|---|---|---|---|---|
| 801 821 | AH1 | 801 829 | AH1 | 801 837 | AH1 | 801 845 | AH1 | 801 853 | AH1 |
| 801 822 | AH1 | 801 830 | AH1 | 801 838 | AH1 | 801 846 | AH1 | 801 854 | AH1 |
| 801 823 | AH1 | 801 831 | AH1 | 801 839 | AH1 | 801 847 | AH1 | 801 855 | AH1 |
| 801 824 | AH1 | 801 832 | AH1 | 801 840 | AH1 | 801 848 | AH1 | 801 856 | AH1 |
| 801 825 | AH1 | 801 833 | AH1 | 801 841 | AH1 | 801 849 | AH1 | 801 857 | AH1 |
| 801 826 | AH1 | 801 834 | AH1 | 801 842 | AH1 | 801 850 | AH1 | 801 858 | AH1 |
| 801 827 | AH1 | 801 835 | AH1 | 801 843 | AH1 | 801 851 | AH1 | 801 859 | AH1 |
| 801 828 | AH1 | 801 836 | AH1 | 801 844 | AH1 | 801 852 | AH1 | 801 860 | AH1 |

## CLASS 802.0                ICE Trailer Second

**Built:** 1990-92 by Duewag/LHB.
**Wheel Arrangement:** 2-2.                    **Seats:** 66S.
**Weight:** 52.80 tonnes.                       **Maximum Speed:** 280 km/h.
**Length:** 26.400 m.

| | | | | | | | | | |
|---|---|---|---|---|---|---|---|---|---|
| 802 001 | AH1 | 802 021 | AH1 | 802 041 | AH1 | 802 061 | AH1 | 802 080 | AH1 |
| 802 002 | AH1 | 802 022 | AH1 | 802 042 | AH1 | 802 062 | AH1 | 802 081 | AH1 |
| 802 003 | AH1 | 802 023 | AH1 | 802 043 | AH1 | 802 063 | AH1 | 802 082 | AH1 |
| 802 004 | AH1 | 802 024 | AH1 | 802 044 | AH1 | 802 064 | AH1 | 802 083 | AH1 |
| 802 005 | AH1 | 802 025 | AH1 | 802 045 | AH1 | 802 065 | AH1 | 802 084 | AH1 |
| 802 006 | AH1 | 802 026 | AH1 | 802 046 | AH1 | 802 066 | AH1 | 802 085 | AH1 |
| 802 007 | AH1 | 802 027 | AH1 | 802 047 | AH1 | 802 067 | AH1 | 802 086 | AH1 |
| 802 008 | AH1 | 802 028 | AH1 | 802 048 | AH1 | 802 068 | AH1 | 802 087 | AH1 |
| 802 009 | AH1 | 802 029 | AH1 | 802 049 | AH1 | 802 069 | AH1 | 802 088 | AH1 |
| 802 010 | AH1 | 802 030 | AH1 | 802 050 | AH1 | 802 070 | AH1 | 802 089 | AH1 |
| 802 011 | AH1 | 802 031 | AH1 | 802 051 | AH1 | 802 071 | AH1 | 802 090 | AH1 |
| 802 012 | AH1 | 802 032 | AH1 | 802 052 | AH1 | 802 072 | AH1 | 802 091 | AH1 |
| 802 013 | AH1 | 802 033 | AH1 | 802 053 | AH1 | 802 073 | AH1 | 802 092 | AH1 |
| 802 014 | AH1 | 802 034 | AH1 | 802 054 | AH1 | 802 074 | AH1 | 802 093 | AH1 |
| 802 015 | AH1 | 802 035 | AH1 | 802 055 | AH1 | 802 075 | AH1 | 802 094 | AH1 |
| 802 016 | AH1 | 802 036 | AH1 | 802 056 | AH1 | 802 076 | AH1 | 802 095 | AH1 |
| 802 017 | AH1 | 802 037 | AH1 | 802 057 | AH1 | 802 077 | AH1 | 802 096 | AH1 |
| 802 018 | AH1 | 802 038 | AH1 | 802 058 | AH1 | 802 078 | AH1 | 802 097 | AH1 |
| 802 019 | AH1 | 802 039 | AH1 | 802 059 | AH1 | 802 079 | AH1 | 802 098 | AH1 |
| 802 020 | AH1 | 802 040 | AH1 | 802 060 | AH1 | | | | |

## CLASS 802.3                ICE Trailer Second

**Built:** 1990-92 by LHB/MBB/MAN.
**Wheel Arrangement:** 2-2.                    **Seats:** 66S.
**Weight:** 52.80 tonnes.                       **Maximum Speed:** 280 km/h.
**Length:** 26.400 m.

| | | | | | | | | | |
|---|---|---|---|---|---|---|---|---|---|
| 802 301 | AH1 | 802 321 | AH1 | 802 341 | AH1 | 802 361 | AH1 | 802 381 | AH1 |
| 802 302 | AH1 | 802 322 | AH1 | 802 342 | AH1 | 802 362 | AH1 | 802 382 | AH1 |
| 802 303 | AH1 | 802 323 | AH1 | 802 343 | AH1 | 802 363 | AH1 | 802 383 | AH1 |
| 802 304 | AH1 | 802 324 | AH1 | 802 344 | AH1 | 802 364 | AH1 | 802 384 | AH1 |
| 802 305 | AH1 | 802 325 | AH1 | 802 345 | AH1 | 802 365 | AH1 | 802 385 | AH1 |
| 802 306 | AH1 | 802 326 | AH1 | 802 346 | AH1 | 802 366 | AH1 | 802 386 | AH1 |
| 802 307 | AH1 | 802 327 | AH1 | 802 347 | AH1 | 802 367 | AH1 | 802 387 | AH1 |
| 802 308 | AH1 | 802 328 | AH1 | 802 348 | AH1 | 802 368 | AH1 | 802 388 | AH1 |
| 802 309 | AH1 | 802 329 | AH1 | 802 349 | AH1 | 802 369 | AH1 | 802 389 | AH1 |
| 802 310 | AH1 | 802 330 | AH1 | 802 350 | AH1 | 802 370 | AH1 | 802 390 | AH1 |
| 802 311 | AH1 | 802 331 | AH1 | 802 351 | AH1 | 802 371 | AH1 | 802 391 | AH1 |
| 802 312 | AH1 | 802 332 | AH1 | 802 352 | AH1 | 802 372 | AH1 | 802 392 | AH1 |
| 802 313 | AH1 | 802 333 | AH1 | 802 353 | AH1 | 802 373 | AH1 | 802 393 | AH1 |
| 802 314 | AH1 | 802 334 | AH1 | 802 354 | AH1 | 802 374 | AH1 | 802 394 | AH1 |
| 802 315 | AH1 | 802 335 | AH1 | 802 355 | AH1 | 802 375 | AH1 | 802 395 | AH1 |
| 802 316 | AH1 | 802 336 | AH1 | 802 356 | AH1 | 802 376 | AH1 | 802 396 | AH1 |
| 802 317 | AH1 | 802 337 | AH1 | 802 357 | AH1 | 802 377 | AH1 | 802 397 | AH1 |
| 802 318 | AH1 | 802 338 | AH1 | 802 358 | AH1 | 802 378 | AH1 | 802 398 | AH1 |
| 802 319 | AH1 | 802 339 | AH1 | 802 359 | AH1 | 802 379 | AH1 | 802 399 | AH1 |
| 802 320 | AH1 | 802 340 | AH1 | 802 360 | AH1 | 802 380 | AH1 | 802 400 | AH1 |

| | | | | | | | | | |
|---|---|---|---|---|---|---|---|---|---|
| 802 401 | AH1 | 802 413 | AH1 | 802 425 | AH1 | 802 437 | AH1 | 802 448 | AH1 |
| 802 402 | AH1 | 802 414 | AH1 | 802 426 | AH1 | 802 438 | AH1 | 802 449 | AH1 |
| 802 403 | AH1 | 802 415 | AH1 | 802 427 | AH1 | 802 439 | AH1 | 802 450 | AH1 |
| 802 404 | AH1 | 802 416 | AH1 | 802 428 | AH1 | 802 440 | AH1 | 802 451 | AH1 |
| 802 405 | AH1 | 802 417 | AH1 | 802 429 | AH1 | 802 441 | AH1 | 802 452 | AH1 |
| 802 406 | AH1 | 802 418 | AH1 | 802 430 | AH1 | 802 442 | AH1 | 802 453 | AH1 |
| 802 407 | AH1 | 802 419 | AH1 | 802 431 | AH1 | 802 443 | AH1 | 802 454 | AH1 |
| 802 408 | AH1 | 802 420 | AH1 | 802 432 | AH1 | 802 444 | AH1 | 802 455 | AH1 |
| 802 409 | AH1 | 802 421 | AH1 | 802 433 | AH1 | 802 445 | AH1 | 802 456 | AH1 |
| 802 410 | AH1 | 802 422 | AH1 | 802 434 | AH1 | 802 446 | AH1 | 802 457 | AH1 |
| 802 411 | AH1 | 802 423 | AH1 | 802 435 | AH1 | 802 447 | AH1 | 802 458 | AH1 |
| 802 412 | AH1 | 802 424 | AH1 | 802 436 | AH1 | | | | |

## CLASS 802.6　　　　　　　　　　　ICE Trailer Second

**Built:** 1990-92 by MBB.
**Wheel Arrangement:** 2-2.　　　　　　　　　**Seats:** 66S.
**Weight:** 52.80 tonnes.　　　　　　　　　　**Maximum Speed:** 280 km/h.
**Length:** 26.400 m.

| | | | | | | | | | |
|---|---|---|---|---|---|---|---|---|---|
| 802 601 | AH1 | 802 613 | AH1 | 802 625 | AH1 | 802 637 | AH1 | 802 649 | AH1 |
| 802 602 | AH1 | 802 614 | AH1 | 802 626 | AH1 | 802 638 | AH1 | 802 650 | AH1 |
| 802 603 | AH1 | 802 615 | AH1 | 802 627 | AH1 | 802 639 | AH1 | 802 651 | AH1 |
| 802 604 | AH1 | 802 616 | AH1 | 802 628 | AH1 | 802 640 | AH1 | 802 652 | AH1 |
| 802 605 | AH1 | 802 617 | AH1 | 802 629 | AH1 | 802 641 | AH1 | 802 653 | AH1 |
| 802 606 | AH1 | 802 618 | AH1 | 802 630 | AH1 | 802 642 | AH1 | 802 654 | AH1 |
| 802 607 | AH1 | 802 619 | AH1 | 802 631 | AH1 | 802 643 | AH1 | 802 655 | AH1 |
| 802 608 | AH1 | 802 620 | AH1 | 802 632 | AH1 | 802 644 | AH1 | 802 656 | AH1 |
| 802 609 | AH1 | 802 621 | AH1 | 802 633 | AH1 | 802 645 | AH1 | 802 657 | AH1 |
| 802 610 | AH1 | 802 622 | AH1 | 802 634 | AH1 | 802 646 | AH1 | 802 658 | AH1 |
| 802 611 | AH1 | 802 623 | AH1 | 802 635 | AH1 | 802 647 | AH1 | 802 659 | AH1 |
| 802 612 | AH1 | 802 624 | AH1 | 802 636 | AH1 | 802 648 | AH1 | 802 660 | AH1 |

## CLASS 802.8　　　　　　　　　　　ICE Trailer Second

**Built:** 1990-92 by MBB/MAN.
**Wheel Arrangement:** 2-2.　　　　　　　　　**Seats:** 66S.
**Weight:** 52.80 tonnes.　　　　　　　　　　**Maximum Speed:** 280 km/h.
**Length:** 26.400 m.

| | | | | | | | | | |
|---|---|---|---|---|---|---|---|---|---|
| 802 801 | AH1 | 802 813 | AH1 | 802 825 | AH1 | 802 837 | AH1 | 802 849 | AH1 |
| 802 802 | AH1 | 802 814 | AH1 | 802 826 | AH1 | 802 838 | AH1 | 802 850 | AH1 |
| 802 803 | AH1 | 802 815 | AH1 | 802 827 | AH1 | 802 839 | AH1 | 802 851 | AH1 |
| 802 804 | AH1 | 802 816 | AH1 | 802 828 | AH1 | 802 840 | AH1 | 802 852 | AH1 |
| 802 805 | AH1 | 802 817 | AH1 | 802 829 | AH1 | 802 841 | AH1 | 802 853 | AH1 |
| 802 806 | AH1 | 802 818 | AH1 | 802 830 | AH1 | 802 842 | AH1 | 802 854 | AH1 |
| 802 807 | AH1 | 802 819 | AH1 | 802 831 | AH1 | 802 843 | AH1 | 802 855 | AH1 |
| 802 808 | AH1 | 802 820 | AH1 | 802 832 | AH1 | 802 844 | AH1 | 802 856 | AH1 |
| 802 809 | AH1 | 802 821 | AH1 | 802 833 | AH1 | 802 845 | AH1 | 802 857 | AH1 |
| 802 810 | AH1 | 802 822 | AH1 | 802 834 | AH1 | 802 846 | AH1 | 802 858 | AH1 |
| 802 811 | AH1 | 802 823 | AH1 | 802 835 | AH1 | 802 847 | AH1 | 802 859 | AH1 |
| 802 812 | AH1 | 802 824 | AH1 | 802 836 | AH1 | 802 848 | AH1 | 802 860 | AH1 |

## CLASS 803　　　　　　　　　ICE Special Trailer Second

**Built:** 1990-92 by Duewag.
**Wheel Arrangement:** 2-2.　　　　　　　　　**Seats:** 39S + 4 seat conference room.
**Weight:** 53.60 tonnes.　　　　　　　　　　**Maximum Speed:** 280 km/h.
**Length:** 26.400 m.

| | | | | | | | | | |
|---|---|---|---|---|---|---|---|---|---|
| 803 001 | AH1 | 803 006 | AH1 | 803 011 | AH1 | 803 016 | AH1 | 803 021 | AH1 |
| 803 002 | AH1 | 803 007 | AH1 | 803 012 | AH1 | 803 017 | AH1 | 803 022 | AH1 |
| 803 003 | AH1 | 803 008 | AH1 | 803 013 | AH1 | 803 018 | AH1 | 803 023 | AH1 |
| 803 004 | AH1 | 803 009 | AH1 | 803 014 | AH1 | 803 019 | AH1 | 803 024 | AH1 |
| 803 005 | AH1 | 803 010 | AH1 | 803 015 | AH1 | 803 020 | AH1 | 803 025 | AH1 |

# GERMANY

| | | | | | | | | | |
|---|---|---|---|---|---|---|---|---|---|
| 803 026 | AH1 | 803 033 | AH1 | 803 040 | AH1 | 803 047 | AH1 | 803 054 | AH1 |
| 803 027 | AH1 | 803 034 | AH1 | 803 041 | AH1 | 803 048 | AH1 | 803 055 | AH1 |
| 803 028 | AH1 | 803 035 | AH1 | 803 042 | AH1 | 803 049 | AH1 | 803 056 | AH1 |
| 803 029 | AH1 | 803 036 | AH1 | 803 043 | AH1 | 803 050 | AH1 | 803 057 | AH1 |
| 803 030 | AH1 | 803 037 | AH1 | 803 044 | AH1 | 803 051 | AH1 | 803 058 | AH1 |
| 803 031 | AH1 | 803 038 | AH1 | 803 045 | AH1 | 803 052 | AH1 | 803 059 | AH1 |
| 803 032 | AH1 | 803 039 | AH1 | 803 046 | AH1 | 803 053 | AH1 | 803 060 | AH1 |

## CLASS 804 — ICE Restaurant/Bar

**Built:** 1990-92 by Waggon Union.
**Wheel Arrangement:** 2-2.
**Weight:** 58.20 tonnes.
**Length:** 26.400 m.
**Seats:** 24 restaurant seats + bar seating.
**Maximum Speed:** 280 km/h.

| | | | | | | | | | |
|---|---|---|---|---|---|---|---|---|---|
| 804 001 | AH1 | 804 013 | AH1 | 804 025 | AH1 | 804 037 | AH1 | 804 049 | AH1 |
| 804 002 | AH1 | 804 014 | AH1 | 804 026 | AH1 | 804 038 | AH1 | 804 050 | AH1 |
| 804 003 | AH1 | 804 015 | AH1 | 804 027 | AH1 | 804 039 | AH1 | 804 051 | AH1 |
| 804 004 | AH1 | 804 016 | AH1 | 804 028 | AH1 | 804 040 | AH1 | 804 052 | AH1 |
| 804 005 | AH1 | 804 017 | AH1 | 804 029 | AH1 | 804 041 | AH1 | 804 053 | AH1 |
| 804 006 | AH1 | 804 018 | AH1 | 804 030 | AH1 | 804 042 | AH1 | 804 054 | AH1 |
| 804 007 | AH1 | 804 019 | AH1 | 804 031 | AH1 | 804 043 | AH1 | 804 055 | AH1 |
| 804 008 | AH1 | 804 020 | AH1 | 804 032 | AH1 | 804 044 | AH1 | 804 056 | AH1 |
| 804 009 | AH1 | 804 021 | AH1 | 804 033 | AH1 | 804 045 | AH1 | 804 057 | AH1 |
| 804 010 | AH1 | 804 022 | AH1 | 804 034 | AH1 | 804 046 | AH1 | 804 058 | AH1 |
| 804 011 | AH1 | 804 023 | AH1 | 804 035 | AH1 | 804 047 | AH1 | 804 059 | AH1 |
| 804 012 | AH1 | 804 024 | AH1 | 804 036 | AH1 | 804 048 | AH1 | 804 060 | AH1 |

## CLASSES 403/404 — 4-Car Unit

**Built:** 1973 by LHB/MAN/MBB.
**Electrical Equipment:** AEG/BBC/Siemens.
**Continuous Rating:** 960 kW per car.
**Wheel Arrangement:** Bo-Bo + Bo-Bo + Bo-Bo + Bo-Bo.
**Weight:** 235.70 tonnes.
**Maximum Speed:** 200 km/h.
**Length:** 27.450 + 27.160 + 27.160 + 27.450 m.
**Seats:** 45F + 51F + 42F + 45F.
**Note:** Do not always operate in fixed formations.

| | | | | | | | | | | |
|---|---|---|---|---|---|---|---|---|---|---|
| 403 001 | 404 001 | 404 101 | 403 002 | (Z) | | 403 005 | 404 003 | 404 103 | 403 006 | (Z) |
| 403 003 | 404 002 | 404 102 | 403 004 | (Z) | | | | | | |

## CLASSES 410 & 810 — IC-Experimental

**Built:** 1986 by Krupp/Henschel (Power Cars), MAN Trailers.
**Electrical Equipment:** Siemens.
**Continuous Rating:** 4200 kW.
**Wheel Arrangement:** Bo-Bo + 2-2 + 2-2 + 2-2 + Bo-Bo.
**Weight:** 78.20 + 51.00 + 51.00 + 51.00 + 78.20 tonnes.
**Length:** 20.810 + 24.340 + 24.340 + 24.340 + 20.810 m.
**Maximum Speed:** 350 km/h.
**Seats:** 0 + 36F + 24F, 27S + 0 + 0.

| | | | | | |
|---|---|---|---|---|---|
| 410 001 | 810 001 | 810 002 | 810 003 | 410 002 | FF1 |

## CLASSES 420/421 — 3-Car Unit

**Built:** 1969-90 by LHB/MBB/O & K/Uerdingen/Waggon Union.
**Electrical Equipment:** AEG/BBC/Siemens.
**Continuous Rating:** 800 kW per car.
**Wheel Arrangements:** Bo-Bo + Bo-Bo + Bo-Bo.
**Weight:** 44.00 + 41.00 + 44.00 tonnes.
**Seats:** 63S + 17F 49S + 65S.
**Length:** 23.300 + 20.800 + 23.300 m.
**Maximum Speed:** 120 km/h.

| | | | | | | | | |
|---|---|---|---|---|---|---|---|---|
| 420 001 | 421 001 | 420 501 | MH6 | | 420 005 | 421 005 | 420 505 | MH6 |
| 420 002 | 421 002 | 420 502 | MH6 | | 420 006 | 421 006 | 420 506 | MH6 |
| 420 003 | 421 003 | 420 503 | MH6 | | 420 007 | 421 007 | 420 507 | MH6 |
| 420 004 | 421 004 | 420 504 | MH6 | | 420 008 | 421 008 | 420 508 | MH6 |

| | | | |
|---|---|---|---|
| 420 009 | 421 009 | 420 509 | MH6 |
| 420 010 | 421 010 | 420 510 | MH6 |
| 420 011 | 421 011 | 420 511 | MH6 |
| 420 012 | 421 012 | 420 512 | MH6 |
| 420 013 | 421 013 | 420 513 | MH6 |
| 420 014 | 421 014 | 420 514 | MH6 |
| 420 015 | 421 015 | 420 515 | MH6 |
| | 421 016 | 420 516 | MH6 |
| 420 017 | 421 017 | 420 517 | MH6 |
| 420 018 | 421 018 | 420 518 | MH6 |
| 420 019 | 421 019 | 420 519 | MH6 |
| 420 020 | 421 020 | 420 520 | MH6 |
| 420 021 | 421 021 | 420 521 | MH6 |
| 420 022 | 421 022 | 420 522 | MH6 |
| 420 023 | 421 023 | 420 523 | MH6 |
| 420 024 | 421 024 | 420 524 | MH6 |
| 420 025 | 421 025 | 420 525 | MH6 |
| 420 026 | 421 026 | 420 526 | MH6 |
| 420 027 | 421 027 | 420 527 | MH6 |
| 420 028 | 421 028 | 420 528 | MH6 |
| 420 029 | 421 029 | 420 529 | MH6 |
| 420 030 | 421 030 | 420 530 | MH6 |
| 420 031 | 421 031 | 420 531 | MH6 |
| 420 032 | 421 032 | 420 532 | MH6 |
| 420 033 | 421 033 | 420 533 | MH6 |
| 420 034 | 421 034 | 420 534 | MH6 |
| 420 035 | 421 035 | 420 535 | MH6 |
| 420 036 | 421 036 | 420 536 | MH6 |
| 420 037 | 421 037 | 420 537 | MH6 |
| 420 038 | 421 038 | 420 538 | MH6 |
| 420 039 | 421 039 | 420 539 | MH6 |
| 420 040 | 421 040 | 420 540 | MH6 |
| 420 041 | 421 041 | 420 541 | MH6 |
| 420 042 | 421 042 | 420 542 | MH6 |
| 420 043 | 421 043 | 420 543 | MH6 |
| 420 044 | 421 044 | 420 544 | MH6 |
| 420 045 | 421 045 | 420 545 | MH6 |
| 420 046 | 421 046 | 420 546 | MH6 |
| 420 047 | 421 047 | 420 547 | MH6 |
| 420 048 | 421 048 | 420 548 | MH6 |
| 420 049 | 421 049 | 420 549 | MH6 |
| 420 050 | 421 050 | 420 550 | MH6 |
| 420 051 | 421 051 | 420 551 | MH6 |
| 420 052 | 421 052 | 420 552 | MH6 |
| 420 053 | 421 053 | 420 553 | MH6 |
| 420 054 | 421 054 | 420 554 | MH6 |
| 420 055 | 421 055 | 420 555 | MH6 |
| 420 056 | 421 056 | 420 556 | MH6 |
| 420 057 | 421 057 | 420 557 | MH6 |
| 420 058 | 421 058 | 420 558 | MH6 |
| 420 059 | 421 059 | 420 559 | MH6 |
| 420 060 | 421 060 | 420 560 | MH6 |
| 420 061 | 421 061 | 420 561 | MH6 |
| 420 062 | 421 062 | 420 562 | MH6 |
| 420 063 | 421 063 | 420 563 | MH6 |
| 420 064 | 421 064 | 420 564 | MH6 |
| 420 065 | 421 065 | 420 565 | MH6 |
| 420 066 | 421 066 | 420 566 | MH6 |
| 420 067 | 421 067 | 420 567 | MH6 |
| 420 068 | 421 068 | 420 568 | MH6 |
| 420 069 | 421 069 | 420 569 | MH6 |
| 420 070 | 421 070 | 420 570 | MH6 |
| 420 071 | 421 071 | 420 571 | MH6 |
| 420 072 | 421 072 | 420 572 | MH6 |
| 420 073 | 421 073 | 420 573 | MH6 |
| 420 074 | 421 074 | 420 574 | MH6 |
| 420 075 | 421 075 | 420 575 | MH6 |
| 420 076 | 421 076 | 420 576 | MH6 |
| 420 077 | 421 077 | 420 577 | MH6 |
| 420 078 | 421 078 | 420 578 | MH6 |
| 420 079 | 421 079 | 420 579 | MH6 |
| 420 080 | 421 080 | 420 580 | MH6 |
| 420 081 | 421 081 | 420 581 | MH6 |
| 420 082 | 421 082 | 420 582 | MH6 |
| 420 083 | 421 083 | 420 583 | MH6 |
| 420 084 | 421 084 | 420 584 | MH6 |
| 420 085 | 421 085 | 420 585 | MH6 |
| 420 086 | 421 086 | 420 586 | MH6 |
| 420 087 | 421 087 | 420 587 | MH6 |
| 420 088 | 421 088 | 420 588 | MH6 |
| 420 089 | 421 089 | 420 589 | MH6 |
| 420 090 | 421 090 | 420 590 | MH6 |
| 420 091 | 421 091 | 420 591 | MH6 |
| 420 092 | 421 092 | 420 592 | MH6 |
| 420 093 | 421 093 | 420 593 | MH6 |
| 420 094 | 421 094 | 420 594 | MH6 |
| 420 095 | 421 095 | 420 595 | MH6 |
| 420 096 | 421 096 | 420 596 | MH6 |
| 420 097 | 421 097 | 420 597 | MH6 |
| 420 098 | 421 098 | 420 598 | MH6 |
| 420 099 | 421 099 | 420 599 | MH6 |
| 420 100 | 421 100 | 420 600 | MH6 |
| 420 101 | 421 101 | 420 601 | MH6 |
| 420 102 | 421 102 | 420 602 | MH6 |
| 420 103 | 421 103 | 420 603 | MH6 |
| 420 104 | 421 104 | 420 604 | MH6 |
| 420 105 | 421 105 | 420 605 | MH6 |
| 420 106 | 421 106 | 420 606 | MH6 |
| 420 107 | 421 107 | 420 607 | MH6 |
| 420 108 | 421 108 | 420 608 | MH6 |
| 420 109 | 421 109 | 420 609 | MH6 |
| 420 110 | 421 110 | 420 610 | MH6 |
| 420 111 | 421 111 | 420 611 | MH6 |
| 420 112 | 421 112 | 420 612 | MH6 |
| 420 113 | 421 113 | 420 613 | MH6 |
| 420 114 | 421 114 | 420 614 | MH6 |
| 420 115 | 421 115 | 420 615 | MH6 |
| 420 116 | 421 116 | 420 616 | MH6 |
| 420 117 | 421 117 | 420 617 | MH6 |
| 420 118 | 421 118 | 420 618 | MH6 |
| 420 119 | 421 119 | 420 619 | MH6 |
| 420 120 | 421 120 | 420 620 | MH6 |
| 420 121 | 421 121 | 420 621 | MH6 |
| 420 122 | 421 122 | 420 622 | MH6 |
| 420 123 | 421 123 | 420 623 | MH6 |
| 420 124 | 421 124 | 420 624 | MH6 |
| 420 125 | 421 125 | 420 625 | MH6 |
| 420 126 | 421 126 | 420 626 | MH6 |
| 420 127 | 421 127 | 420 627 | MH6 |
| 420 128 | 421 128 | 420 628 | MH6 |
| 420 129 | 421 129 | 420 629 | MH6 |
| 420 130 | 421 130 | 420 630 | MH6 |
| 420 131 | 421 131 | 420 631 | MH6 |
| 420 132 | 421 132 | 420 632 | MH6 |
| 420 133 | 421 133 | 420 633 | MH6 |
| 420 134 | 421 134 | 420 634 | MH6 |

# GERMANY

| | | | | | | | |
|---|---|---|---|---|---|---|---|
| 420 135 | 421 135 | 420 635 | MH6 | 420 198 | 421 198 | 420 698 | MH6 |
| 420 136 | 421 136 | 420 636 | MH6 | 420 199 | 421 199 | 420 699 | FFG |
| 420 137 | 421 137 | 420 637 | MH6 | 420 200 | 421 200 | 420 700 | FFG |
| 420 138 | 421 138 | 420 638 | MH6 | 420 201 | 421 201 | 420 701 | FFG |
| 420 139 | 421 139 | 420 639 | MH6 | 420 202 | 421 202 | 420 702 | FFG |
| 420 140 | 421 140 | 420 640 | MH6 | 420 203 | 421 203 | 420 703 | FFG |
| 420 141 | 421 141 | 420 641 | MH6 | 420 204 | 421 204 | 420 704 | FFG |
| 420 142 | 421 142 | 420 642 | MH6 | 420 205 | 421 205 | 420 705 | FFG |
| 420 143 | 421 143 | 420 643 | MH6 | 420 206 | 421 206 | 420 706 | FFG |
| 420 144 | 421 144 | 420 644 | MH6 | 420 207 | 421 207 | 420 707 | FFG |
| 420 145 | 421 145 | 420 645 | MH6 | 420 208 | 421 208 | 420 708 | FFG |
| 420 146 | 421 146 | 420 646 | MH6 | 420 209 | 421 209 | 420 709 | FFG |
| 420 147 | 421 147 | 420 647 | MH6 | 420 211 | 421 211 | 420 711 | FFG |
| 420 148 | 421 148 | 420 648 | MH6 | 420 212 | 421 212 | 420 712 | FFG |
| 420 149 | 421 149 | 420 649 | MH6 | 420 213 | 421 213 | 420 713 | FFG |
| 420 150 | 421 150 | 420 650 | MH6 | 420 214 | 421 214 | 420 714 | FFG |
| 420 151 | 421 151 | 420 651 | MH6 | 420 215 | 421 215 | 420 715 | FFG |
| 420 152 | 421 152 | 420 652 | MH6 | 420 216 | 421 216 | 420 716 | FFG |
| 420 153 | 421 153 | 420 653 | MH6 | 420 217 | 421 217 | 420 717 | FFG |
| 420 154 | 421 154 | 420 654 | MH6 | 420 218 | 421 218 | 420 718 | FFG |
| 420 155 | 421 155 | 420 655 | MH6 | 420 219 | 421 219 | 420 719 | FFG |
| 420 156 | 421 156 | 420 656 | MH6 | 420 220 | 421 220 | 420 720 | FFG |
| 420 157 | 421 157 | 420 657 | MH6 | 420 221 | 421 221 | 420 721 | FFG |
| 420 158 | 421 158 | 420 658 | MH6 | 420 222 | 421 222 | 420 722 | FFG |
| 420 159 | 421 159 | 420 659 | MH6 | 420 223 | 421 223 | 420 723 | FFG |
| 420 160 | 421 160 | | MH6 | 420 224 | 421 224 | 420 724 | FFG |
| 420 161 | 421 161 | 420 661 | MH6 | 420 225 | 421 225 | 420 725 | FFG |
| 420 162 | 421 162 | 420 662 | MH6 | 420 226 | 421 226 | 420 726 | FFG |
| 420 163 | 421 163 | 420 663 | MH6 | 420 227 | 421 227 | 420 727 | FFG |
| 420 164 | 421 164 | 420 664 | MH6 | 420 228 | 421 228 | 420 728 | FFG |
| 420 165 | 421 165 | 420 665 | MH6 | 420 229 | 421 229 | 420 729 | FFG |
| 420 166 | 421 166 | 420 666 | MH6 | 420 230 | 421 230 | 420 730 | FFG |
| 420 167 | 421 167 | 420 667 | MH6 | 420 231 | 421 231 | 420 731 | FFG |
| 420 168 | 421 168 | 420 668 | MH6 | 420 232 | 421 232 | 420 732 | FFG |
| 420 169 | 421 169 | 420 669 | MH6 | 420 233 | 421 233 | 420 733 | FFG |
| 420 170 | 421 170 | 420 670 | MH6 | 420 234 | 421 234 | 420 734 | FFG |
| 420 171 | 421 171 | 420 671 | MH6 | 420 235 | 421 235 | 420 735 | FFG |
| 420 172 | 421 172 | 420 672 | MH6 | 420 236 | 421 236 | 420 736 | FFG |
| 420 173 | 421 173 | 420 673 | MH6 | 420 237 | 421 237 | 420 737 | FFG |
| 420 174 | 421 174 | 420 674 | MH6 | 420 238 | 421 238 | 420 738 | FFG |
| 420 175 | 421 175 | 420 675 | MH6 | 420 239 | 421 239 | 420 739 | FFG |
| 420 176 | 421 176 | 420 676 | MH6 | 420 240 | 421 240 | 420 740 | FFG |
| 420 177 | 421 177 | 420 677 | MH6 | 420 241 | 421 241 | 420 741 | FFG |
| 420 178 | 421 178 | 420 678 | MH6 | 420 242 | 421 242 | 420 742 | FFG |
| 420 179 | 421 179 | 420 679 | MH6 | 420 243 | 421 243 | 420 743 | FFG |
| 420 180 | 421 180 | 420 680 | MH6 | 420 244 | 421 244 | 420 744 | FFG |
| 420 181 | 421 181 | 420 681 | MH6 | 420 245 | 421 245 | 420 745 | FFG |
| 420 182 | 421 182 | 420 682 | MH6 | 420 246 | 421 246 | 420 746 | FFG |
| 420 183 | 421 183 | 420 683 | MH6 | 420 247 | 421 247 | 420 747 | FFG |
| 420 184 | 421 184 | 420 684 | MH6 | 420 248 | 421 248 | 420 748 | FFG |
| 420 185 | 421 185 | 420 685 | MH6 | 420 249 | 421 249 | 420 749 | FFG |
| 420 186 | 421 186 | 420 686 | MH6 | 420 250 | 421 250 | 420 750 | FFG |
| 420 187 | 421 187 | 420 687 | MH6 | 420 251 | 421 251 | 420 751 | FFG |
| 420 188 | 421 188 | 420 688 | MH6 | 420 252 | 421 252 | 420 752 | FFG |
| 420 189 | 421 189 | 420 689 | MH6 | 420 253 | 421 253 | 420 753 | FFG |
| 420 190 | 421 190 | 420 690 | MH6 | 420 254 | 421 254 | 420 754 | FFG |
| 420 191 | 421 191 | 420 691 | MH6 | 420 255 | 421 255 | 420 755 | FFG |
| 420 192 | 421 192 | 420 692 | MH6 | 420 256 | 421 256 | 420 756 | FFG |
| 420 193 | 421 193 | 420 693 | MH6 | 420 257 | 421 257 | 420 757 | FFG |
| 420 194 | 421 194 | 420 694 | MH6 | 420 258 | 421 258 | 420 758 | FFG |
| 420 195 | 421 195 | 420 695 | MH6 | 420 259 | 421 259 | 420 759 | FFG |
| 420 196 | 421 196 | 420 696 | MH6 | 420 260 | 421 260 | 420 760 | FFG |
| 420 197 | 421 197 | 420 697 | MH6 | 420 261 | 421 261 | 420 761 | FFG |

| | | | | | | | |
|---|---|---|---|---|---|---|---|
| 420 262 | 421 262 | 420 762 | FFG | 420 325 | 421 325 | 420 825 | TP |
| 420 263 | 421 263 | 420 763 | FFG | 420 326 | 421 326 | 420 826 | TP |
| 420 264 | 421 264 | 420 764 | FFG | 420 327 | 421 327 | 420 827 | TP |
| 420 265 | 421 265 | 420 765 | FFG | 420 328 | 421 328 | 420 828 | TP |
| 420 266 | 421 266 | 420 766 | FFG | 420 329 | 421 329 | 420 829 | TP |
| 420 267 | 421 267 | 420 767 | FFG | 420 330 | 421 330 | 420 830 | TP |
| 420 268 | 421 268 | 420 768 | FFG | 420 331 | 421 331 | 420 831 | TP |
| 420 269 | 421 269 | 420 769 | FFG | 420 332 | 421 332 | 420 832 | TP |
| 420 270 | 421 270 | 420 770 | FFG | 420 333 | 421 333 | 420 833 | TP |
| 420 271 | 421 271 | 420 771 | FFG | 420 334 | 421 334 | 420 834 | TP |
| 420 272 | 421 272 | 420 772 | FFG | 420 335 | 421 335 | 420 835 | TP |
| 420 273 | 421 273 | 420 773 | FFG | 420 336 | 421 336 | 420 836 | TP |
| 420 274 | 421 274 | 420 774 | FFG | 420 337 | 421 337 | 420 837 | TP |
| 420 275 | 421 275 | 420 775 | FFG | 420 338 | 421 338 | 420 838 | TP |
| 420 276 | 421 276 | 420 776 | FFG | 420 339 | 421 339 | 420 839 | TP |
| 420 277 | 421 277 | 420 777 | FFG | 420 340 | 421 340 | 420 840 | TP |
| 420 278 | 421 278 | 420 778 | FFG | 420 341 | 421 341 | 420 841 | TP |
| 420 279 | | 420 779 | (Z) | 420 342 | 421 342 | 420 842 | TP |
| 420 280 | 421 280 | 420 780 | FFG | 420 343 | 421 343 | 420 843 | TP |
| 420 281 | 421 281 | 420 781 | FFG | 420 344 | 421 344 | 420 844 | TP |
| 420 282 | 421 282 | 420 782 | FFG | 420 345 | 421 345 | 420 845 | TP |
| 420 283 | 421 283 | 420 783 | FFG | 420 346 | 421 346 | 420 846 | TP |
| 420 284 | 421 284 | 420 784 | FFG | 420 347 | 421 347 | 420 847 | TP |
| 420 285 | 421 285 | 420 785 | FFG | 420 348 | 421 348 | 420 848 | TP |
| 420 286 | 421 286 | 420 786 | FFG | 420 349 | 421 349 | 420 849 | TP |
| 420 287 | 421 287 | 420 787 | FFG | 420 350 | 421 350 | 420 850 | TP |
| 420 288 | 421 288 | 420 788 | FFG | 420 351 | 421 351 | 420 851 | TP |
| 420 289 | 421 289 | 420 789 | FFG | 420 352 | 421 352 | 420 852 | TP |
| 420 290 | 421 290 | 420 790 | FFG | 420 353 | 421 353 | 420 853 | TP |
| 420 291 | 421 291 | 420 791 | FFG | 420 354 | 421 354 | 420 854 | TP |
| 420 292 | 421 292 | 420 792 | FFG | 420 355 | 421 355 | 420 855 | TP |
| 420 293 | 421 293 | 420 793 | FFG | 420 356 | 421 356 | 420 856 | TP |
| 420 294 | 421 294 | 420 794 | FFG | 420 357 | 421 357 | 420 857 | TP |
| 420 295 | 421 295 | 420 795 | FFG | 420 358 | 421 358 | 420 858 | TP |
| 420 296 | 421 296 | 420 796 | FFG | 420 359 | 421 359 | 420 859 | TP |
| 420 297 | 421 297 | 420 797 | FFG | 420 360 | 421 360 | 420 860 | TP |
| 420 298 | 421 298 | 420 798 | FFG | 420 361 | 421 361 | 420 861 | TP |
| 420 299 | 421 299 | 420 799 | FFG | 420 362 | 421 362 | 420 862 | TP |
| 420 300 | 421 300 | 420 800 | FFG | 420 363 | 421 363 | 420 863 | TP |
| 420 301 | 421 301 | 420 801 | FFG | 420 364 | 421 364 | 420 864 | TP |
| 420 302 | 421 302 | 420 802 | KD | 420 365 | 421 365 | 420 865 | TP |
| 420 303 | 421 303 | 420 803 | KD | 420 366 | 421 366 | 420 866 | TP |
| 420 304 | 421 304 | 420 804 | KD | 420 367 | 421 367 | 420 867 | TP |
| 420 305 | 421 305 | 420 805 | KD | 420 368 | 421 368 | 420 868 | TP |
| 420 306 | 421 306 | 420 806 | FFG | 420 369 | 421 369 | 420 869 | TP |
| 420 307 | 421 307 | 420 807 | FFG | 420 370 | 421 370 | 420 870 | TP |
| 420 308 | 421 308 | 420 808 | KD | 420 371 | 421 371 | 420 871 | TP |
| 420 309 | 421 309 | 420 809 | KD | 420 372 | 421 372 | 420 872 | TP |
| 420 310 | 421 310 | 420 810 | KD | 420 373 | 421 373 | 420 873 | TP |
| 420 311 | 421 311 | 420 811 | KD | 420 374 | 421 374 | 420 874 | TP |
| 420 312 | 421 312 | 420 812 | KD | 420 375 | 421 375 | 420 875 | TP |
| 420 313 | 421 313 | 420 813 | KD | 420 376 | 421 376 | 420 876 | TP |
| 420 314 | 421 314 | 420 814 | KD | 420 377 | 421 377 | 420 877 | TP |
| 420 315 | 421 315 | 420 815 | KD | 420 378 | 421 378 | 420 878 | TP |
| 420 316 | 421 316 | 420 816 | FFG | 420 379 | 421 379 | 420 879 | TP |
| 420 317 | 421 317 | 420 817 | KD | 420 380 | 421 380 | 420 880 | TP |
| 420 318 | 421 318 | 420 818 | TP | 420 381 | 421 381 | 420 881 | TP |
| 420 319 | 421 319 | 420 819 | KD | 420 382 | 421 382 | 420 882 | TP |
| 420 320 | 421 320 | 420 820 | TP | 420 383 | 421 383 | 420 883 | TP |
| 420 321 | 421 321 | 420 821 | FFG | 420 384 | 421 384 | 420 884 | TP |
| 420 322 | 421 322 | 420 822 | TP | 420 385 | 421 385 | 420 885 | TP |
| 420 323 | 421 323 | 420 823 | FFG | 420 386 | 421 386 | 420 886 | TP |
| 420 324 | 421 324 | 420 824 | TP | 420 387 | 421 387 | 420 887 | TP |

# GERMANY

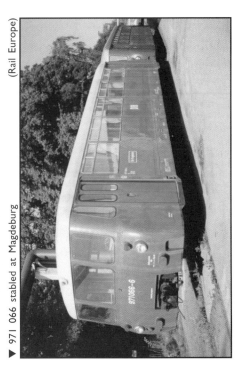

(J. Hayes)

(Rail Europe)

▲ 928 576 and 628 576 at Burghausen on 24.12.94

▼ 971 066 stabled at Magdeburg

(J. Hayes)

(J. Hayes)

▲ A Class 420 3-car emu at Wolfratshausen on 15.05.93

▼ 624 622 heads a 3-car dmu at Coesfeld on 26.05.95

| | | | |
|---|---|---|---|
| 420 388 | 421 388 | 420 888 | TP |
| 420 389 | 421 389 | 420 889 | TP |
| 420 390 | 421 390 | 420 890 | TP |
| 420 400 | 421 400 | 420 900 | TP |
| 420 401 | 421 401 | 420 901 | TP |
| 420 402 | 421 402 | 420 902 | TP |
| 420 403 | 421 403 | 420 903 | TP |
| 420 404 | 421 404 | 420 904 | TP |
| 420 405 | 421 405 | 420 905 | TP |
| 420 406 | 421 406 | 420 906 | TP |
| 420 407 | 421 407 | 420 907 | TP |
| 420 408 | 421 408 | 420 908 | TP |
| 420 409 | 421 409 | 420 909 | TP |
| 420 410 | 421 410 | 420 910 | TP |
| 420 411 | 421 411 | 420 911 | TP |
| 420 412 | 421 412 | 420 912 | TP |
| 420 413 | 421 413 | 420 913 | TP |
| 420 414 | 421 414 | 420 914 | TP |
| 420 415 | 421 415 | 420 915 | TP |
| 420 416 | 421 416 | 420 916 | TP |
| 420 417 | 421 417 | 420 917 | TP |
| 420 418 | 421 418 | 420 918 | TP |
| 420 419 | 421 419 | 420 919 | TP |
| 420 420 | 421 420 | 420 920 | TP |
| 420 421 | 421 421 | 420 921 | TP |
| 420 422 | 421 422 | 420 922 | TP |
| 420 423 | 421 423 | 420 923 | TP |
| 420 424 | 421 424 | 420 924 | TP |
| 420 425 | 421 425 | 420 925 | MH6 |
| 420 426 | 421 426 | 420 926 | MH6 |
| 420 427 | 421 427 | 420 927 | MH6 |
| 420 428 | 421 428 | 420 928 | MH6 |
| 420 429 | 421 429 | 420 929 | MH6 |
| 420 430 | 421 430 | 420 930 | MH6 |
| 420 431 | 421 431 | 420 931 | TP |
| 420 432 | 421 432 | 420 932 | TP |
| 420 433 | 421 433 | 420 933 | TP |
| 420 434 | 421 434 | 420 934 | TP |
| 420 435 | 421 435 | 420 935 | TP |
| 420 436 | 421 436 | 420 936 | TP |
| 420 437 | 421 437 | 420 937 | TP |

| | | | |
|---|---|---|---|
| 420 438 | 421 438 | 420 938 | TP |
| 420 439 | 421 439 | 420 939 | TP |
| 420 440 | 421 440 | 420 940 | TP |
| 420 441 | 421 441 | 420 941 | TP |
| 420 442 | 421 442 | 420 942 | TP |
| 420 443 | 421 443 | 420 943 | TP |
| 420 444 | 421 444 | 420 944 | TP |
| 420 445 | 421 445 | 420 945 | TP |
| 420 446 | 421 446 | 420 946 | TP |
| 420 447 | 421 447 | 420 947 | TP |
| 420 448 | 421 448 | 420 948 | TP |
| 420 449 | 421 449 | 420 949 | TP |
| 420 450 | 421 450 | 420 950 | TP |
| 420 451 | 421 451 | 420 951 | TP |
| 420 452 | 421 452 | 420 952 | TP |
| 420 453 | 421 453 | 420 953 | TP |
| 420 454 | 421 454 | 420 954 | TP |
| 420 455 | 421 455 | 420 955 | TP |
| 420 456 | 421 456 | 420 956 | TP |
| 420 457 | 421 457 | 420 957 | TP |
| 420 458 | 421 458 | 420 958 | TP |
| 420 459 | 421 459 | 420 959 | TP |
| 420 460 | 421 460 | 420 960 | TP |
| 420 461 | 421 461 | 420 961 | TP |
| 420 462 | 421 462 | 420 962 | TP |
| 420 463 | 421 463 | 420 963 | TP |
| 420 464 | 421 464 | 420 964 | TP |
| 420 465 | 421 465 | 420 965 | TP |
| 420 466 | 421 466 | 420 966 | TP |
| 420 467 | 421 467 | 420 967 | TP |
| 420 468 | 421 468 | 420 968 | |
| 420 469 | 421 469 | 420 969 | |
| 420 470 | 421 470 | 420 970 | |
| 420 471 | 421 471 | 420 971 | |
| 420 472 | 421 472 | 420 972 | |
| 420 473 | 421 473 | 420 973 | |
| 420 474 | 421 474 | 420 974 | |
| 420 475 | 421 475 | 420 975 | |
| 420 476 | 421 476 | 420 976 | |
| 420 477 | 421 477 | 420 977 | |

**Names:**

| | |
|---|---|
| 420 217 | Friedberg |
| 420 221 | Raunheim |
| 420 234 | Mainz Kastel |
| 420 236 | Wiesbaden |
| 420 246 | Niedernhausen |
| 420 251 | Kronberg |
| 420 252 | Friedrichsdorf |
| 420 253 | Bad Soden |
| 420 270 | Schwalbach |
| 420 271 | Hofheim am Taunus |
| 420 272 | Hochheim |
| 420 273 | Bad Homberg |
| 420 274 | Bad Vibel |
| 420 275 | Flörsheim am Main |
| 420 276 | Eschborn |
| 420 277 | Sulzbach |
| 420 278 | Oberursel |
| 420 279 | Frankfurt am Main |
| 420 280 | Karben |
| 420 281 | Eppstein |

| | |
|---|---|
| 420 282 | Kriftel |
| 420 283 | Lorsbach |
| 420 284 | Höchst |
| 420 285 | Mainz |
| 420 286 | Niederjosbach |
| 420 287 | Dortelweil |
| 420 288 | Ludwigsburg |
| 420 305 | Stadt Plochingen |
| 420 320 | Plochingen |
| 420 331 | Kreis Ludwigsburg |
| 420 332 | Landeshauptstadt Stuttgart |
| 420 333 | Waiblingen |
| 420 334 | Landkreis Esslingen |
| 420 335 | Rems-Murr Kreis |
| 420 337 | Untertürkheim |
| 420 338 | Böblingen |
| 420 383 | Backnang |
| 420 385 | Bietigheim-Bissingen |
| 420 389 | Bad Canstatt |
| 420 416 | Leonberg |

# GERMANY

## CLASS 450         Karlsruhe Stadtbahn Tram

**Built:** 1994 by
**Supply System:** 750 V dc overhead or 15 kV 16.67 Hz ac overhead.
**Electrical Equipment:**            **Weight:**
**Continuous Rating:**              **Length:**
**Wheel Arrangement:**            **Maximum Speed:**
**Note:** Used in a common pool with other Albtal-Verkerhs-Gesellschaft trams. Also carry local numbers 817-820.

| | | | |
|---|---|---|---|
| 450 001 RK | 450 002 RK | 450 003 RK | 450 004 RK |

## CLASSES 470/870            3-Car Unit

**Built:** 1959-70 by MAN/Wegmann.
**Supply System:** 1200 V dc side contact third rail.
**Electrical Equipment:** BBC/Siemens.
**Continuous Rating:** 640 kW per power car.     **Seats:** 63S + 66F + 63S.
**Wheel Arrangements:** Bo-Bo + 2-2 + Bo-Bo.     **Length:** 22.280 + 20.960 + 22.280 m.
**Weight:** 42.00 + 27.00 + 42.00 tonnes.     **Maximum Speed:** 80 km/h.

| | | | | | | | |
|---|---|---|---|---|---|---|---|
| 470 101 | 870 101 | 470 401 | AOP | 470 124 | 870 124 | 470 424 | AOP |
| 470 102 | 870 102 | 470 402 | AOP | 470 125 | 870 125 | 470 425 | AOP |
| 470 103 | 870 103 | 470 403 | AOP | 470 126 | 870 126 | 470 426 | AOP |
| 470 104 | 870 104 | 470 404 | AOP | 470 127 | 870 127 | 470 427 | AOP |
| 470 105 | 870 105 | 470 405 | AOP | 470 128 | 870 128 | 470 428 | AOP |
| 470 106 | 870 106 | 470 406 | AOP | 470 129 | 870 129 | 470 429 | AOP |
| 470 107 | 870 107 | 470 407 | AOP | 470 130 | 870 130 | 470 430 | AOP |
| 470 108 | 870 108 | 470 408 | AOP | 470 131 | 870 131 | 470 431 | AOP |
| 470 109 | 870 109 | 470 409 | AOP | 470 132 | 870 132 | 470 432 | AOP |
| 470 110 | 870 110 | 470 410 | AOP | 470 133 | 870 133 | 470 433 | AOP |
| 470 111 | 870 111 | 470 411 | AOP | 470 134 | 870 134 | 470 434 | AOP |
| 470 112 | 870 112 | 470 412 | AOP | 470 135 | 870 135 | 470 435 | AOP |
| 470 113 | 870 113 | 470 413 | AOP | 470 136 | 870 136 | 470 436 | AOP |
| 470 114 | 870 114 | 470 414 | AOP | 470 137 | 870 137 | 470 437 | AOP |
| 470 115 | 870 115 | 470 415 | AOP | 470 138 | 870 138 | 470 438 | AOP |
| 470 116 | 870 116 | 470 416 | AOP | 470 139 | 870 139 | 470 439 | AOP |
| 470 117 | 870 117 | 470 417 | AOP | 470 140 | 870 140 | 470 440 | AOP |
| 470 118 | 870 118 | 470 418 | AOP | 470 141 | 870 141 | 470 441 | AOP |
| 470 119 | 870 119 | 470 419 | AOP | 470 142 | 870 142 | 470 442 | AOP |
| 470 120 | 870 120 | 470 420 | AOP | 470 143 | 870 143 | 470 443 | AOP |
| 470 121 | 870 121 | 470 421 | AOP | 470 144 | 870 144 | 470 444 | AOP |
| 470 122 | 870 122 | 470 422 | AOP | 470 145 | 870 145 | 470 445 | AOP |
| 470 123 | 870 123 | 470 423 | AOP | | | | |

## CLASSES 471/871            3-Car Unit

**Built:** 1939-58 by MAN/LHW/Wegmann.
**Supply System:** 1200 V dc side contact third rail.
**Electrical Equipment:** BBC.         **Continuous Rating:** 580 kW per power car.
**Wheel Arrangements:** Bo-Bo + 2-2 + Bo-Bo.     **Seats:** 66S + 68F + 66S.
**Weight:** 50.00 + 31.20 + 50.00 tonnes.     **Maximum Speed:** 80 km/h.
**Length:** 21.280 + 19.960 + 21.280 m.

| | | | | | | | |
|---|---|---|---|---|---|---|---|
| 471 101 | 871 101 | 471 401 | AOP | 471 114 | 871 114 | 471 414 | AOP |
| 471 102 | 871 102 | 471 402 | AOP | 471 115 | 871 115 | 471 415 | AOP |
| 471 104 | 871 104 | 471 404 | AOP | 471 116 | 871 116 | 471 416 | AOP |
| 471 105 | 871 105 | 471 405 | AOP | 471 117 | 871 117 | 471 417 | AOP |
| 471 106 | 871 106 | 471 406 | AOP | 471 118 | 871 118 | 471 418 | AOP |
| 471 107 | 871 107 | 471 407 | AOP | 471 120 | 871 120 | 471 420 | AOP |
| 471 109 | 871 109 | 471 409 | AOP | 471 121 | 871 121 | 471 421 | AOP |
| 471 110 | 871 110 | 471 410 | AOP | 471 122 | 871 122 | 471 422 | AOP |
| 471 111 | 871 111 | 471 411 | AOP | 471 123 | 871 123 | 471 423 | AOP |
| 471 112 | 871 112 | 471 412 | AOP | 471 124 | 871 124 | 471 424 | AOP |
| 471 113 | 871 113 | 471 413 | AOP | 471 126 | 871 126 | 471 426 | AOP |

| | | | |
|---|---|---|---|
| 471 127 | 871 127 | 471 427 | AOP |
| 471 128 | 871 128 | 471 428 | AOP |
| 471 129 | 871 129 | 471 429 | AOP |
| 471 130 | 871 130 | 471 430 | AOP |
| 471 131 | 871 131 | 471 431 | AOP |
| 471 132 | 871 132 | 471 432 | AOP |
| 471 134 | 871 134 | 471 434 | AOP |
| 471 135 | 871 135 | 471 435 | AOP |
| 471 137 | 871 137 | 471 437 | AOP |
| 471 138 | 871 138 | 471 438 | AOP |
| 471 140 | 871 140 | 471 440 | AOP |
| 471 142 | 871 142 | 471 442 | AOP |
| 471 143 | 871 143 | 471 443 | AOP |
| 471 151 | 871 151 | 471 451 | AOP |
| 471 152 | 871 152 | 471 452 | AOP |
| 471 161 | 871 161 | 471 461 | AOP |
| 471 162 | 871 162 | 471 462 | AOP |
| 471 163 | 871 163 | 471 463 | AOP |
| 471 164 | 871 164 | 471 464 | AOP |
| 471 165 | 871 165 | 471 465 | AOP |

| | | | |
|---|---|---|---|
| 471 166 | 871 166 | 471 466 | AOP |
| 471 167 | 871 167 | 471 467 | AOP |
| 471 168 | 871 168 | 471 468 | AOP |
| 471 169 | 871 169 | 471 469 | AOP |
| 471 170 | 871 170 | 471 470 | AOP |
| 471 171 | 871 171 | 471 471 | AOP |
| 471 172 | 871 172 | 471 472 | AOP |
| 471 173 | 871 173 | 471 473 | AOP |
| 471 174 | 871 174 | 471 474 | AOP |
| 471 175 | 871 175 | 471 475 | AOP |
| 471 176 | 871 176 | 471 476 | AOP |
| 471 177 | 871 177 | 471 477 | AOP |
| 471 178 | 871 178 | 471 478 | AOP |
| 471 180 | 871 180 | 471 480 | AOP |
| 471 181 | 871 181 | 471 481 | AOP |
| 471 182 | 871 182 | 471 482 | AOP |
| 471 183 | 871 183 | 471 483 | AOP |
| 471 184 | 871 184 | 471 484 | AOP |
| 471 185 | 871 185 | 471 485 | AOP |
| 471 186 | 871 186 | 471 486 | AOP |

**Name:**
471 163  *Wedel*

## CLASSES 472/473                     3-Car Unit

**Built:** 1974-84 by LHW/MBB.
**Supply System:** 1200 V dc side contact third rail.
**Electrical Equipment:** BBC/Siemens.
**Wheel Arrangements:** Bo-Bo + Bo-Bo + Bo-Bo.
**Weight:** 40.00 + 34.00 + 40.00 tonnes.
**Length:** 65.820 m.

**Continuous Rating:** 500 kW per power car.
**Seats:** 65S + 66F + 65S.
**Maximum Speed:** 100 km/h.

| | | | |
|---|---|---|---|
| 472 001 | 473 001 | 472 501 | AOP |
| 472 002 | 473 002 | 472 502 | AOP |
| 472 003 | 473 003 | 472 503 | AOP |
| 472 004 | 473 004 | 472 504 | AOP |
| 472 005 | 473 005 | 472 505 | AOP |
| 472 006 | 473 006 | 472 506 | AOP |
| 472 007 | 473 007 | 472 507 | AOP |
| 472 008 | 473 008 | 472 508 | AOP |
| 472 009 | 473 009 | 472 509 | AOP |
| 472 010 | 473 010 | 472 510 | AOP |
| 472 011 | 473 011 | 472 511 | AOP |
| 472 012 | 473 012 | 472 512 | AOP |
| 472 013 | 473 013 | 472 513 | AOP |
| 472 014 | 473 014 | 472 514 | AOP |
| 472 015 | 473 015 | 472 515 | AOP |
| 472 016 | 473 016 | 472 516 | AOP |
| 472 017 | 473 017 | 472 517 | AOP |
| 472 018 | 473 018 | 472 518 | AOP |
| 472 019 | 473 019 | 472 519 | AOP |
| 472 020 | 473 020 | 472 520 | AOP |
| 472 021 | 473 021 | 472 521 | AOP |
| 472 022 | 473 022 | 472 522 | AOP |
| 472 023 | 473 023 | 472 523 | AOP |
| 472 024 | 473 024 | 472 524 | AOP |
| 472 025 | 473 025 | 472 525 | AOP |
| 472 026 | 473 026 | 472 526 | AOP |
| 472 027 | 473 027 | 472 527 | AOP |
| 472 028 | 473 028 | 472 528 | AOP |
| 472 029 | 473 029 | 472 529 | AOP |
| 472 030 | 473 030 | 472 530 | AOP |
| 472 031 | 473 031 | 472 531 | AOP |
| 472 032 | 473 032 | 472 532 | AOP |

| | | | |
|---|---|---|---|
| 472 033 | 473 033 | 472 533 | AOP |
| 472 034 | 473 034 | 472 534 | AOP |
| 472 035 | 473 035 | 472 535 | AOP |
| 472 036 | 473 036 | 472 536 | AOP |
| 472 037 | 473 037 | 472 537 | AOP |
| 472 038 | 473 038 | 472 538 | AOP |
| 472 039 | 473 039 | 472 539 | AOP |
| 472 040 | 473 040 | 472 540 | AOP |
| 472 041 | 473 041 | 472 541 | AOP |
| 472 042 | 473 042 | 472 542 | AOP |
| 472 043 | 473 043 | 472 543 | AOP |
| 472 044 | 473 044 | 472 544 | AOP |
| 472 045 | 473 045 | 472 545 | AOP |
| 472 046 | 473 046 | 472 546 | AOP |
| 472 047 | 473 047 | 472 547 | AOP |
| 472 048 | 473 048 | 472 548 | AOP |
| 472 049 | 473 049 | 472 549 | AOP |
| 472 050 | 473 050 | 472 550 | AOP |
| 472 051 | 473 051 | 472 551 | AOP |
| 472 052 | 473 052 | 472 552 | AOP |
| 472 053 | 473 053 | 472 553 | AOP |
| 472 054 | 473 054 | 472 554 | AOP |
| 472 055 | 473 055 | 472 555 | AOP |
| 472 056 | 473 056 | 472 556 | AOP |
| 472 057 | 473 057 | 472 557 | AOP |
| 472 058 | 473 058 | 472 558 | AOP |
| 472 059 | 473 059 | 472 559 | AOP |
| 472 060 | 473 060 | 472 560 | AOP |
| 472 061 | 473 061 | 472 561 | AOP |
| 472 062 | 473 062 | 472 562 | AOP |

# GERMANY

**Names:**

| | | | | |
|---|---|---|---|---|
| 472 001 | Alster | | 472 045 | Harburg |
| 472 030 | Elbe | | 472 060 | Süderelbe |

## CLASS 479.2     Single Unit

**Built:** 1923. Rebuilt 1981 by DR Schöneweide Works.
**Supply System:** 500 V dc overhead.
**Electrical Equipment:**     **Continuous Rating:** 120 kW.
**Wheel Arrangement:** Bo.     **Seats:** 24.
**Weight:** 14.40 tonnes.     **Maximum Speed:** 50 km/h.
**Length:** 11.500 m.
**Note:** Previously DR Class 279.2. These vehicles work between Lichtenhain (an der Berg) and Cursdorf.

| | | |
|---|---|---|
| 479 201 US | 479 203 US | 479 205 US |

## CLASSES 491     Single Unit

**Built:** 1936 by Fuchs.
**Electrical Equipment:** AEG.     **Continuous Rating:** 350 kW.
**Wheel Arrangement:** Bo-2.     **Seats:** 72.
**Weight:** 45.40 tonnes.     **Maximum Speed:** 120 km/h.
**Length:** 20.600 m.

491 001 MH1

# BATTERY ELECTRIC MULTIPLE UNITS

## CLASS 515     Single Unit

**Built:** 1959-63 by DWM/MAN/O & K/Rathgeber.
**Batteries:** 520 kWh.
**Electrical Equipment:** AFA/DWM/Schaltbau/Siemens.     **Continuous Rating:** 200 kW.
**Wheel Arrangement:** Bo-2.     **Seats:** 6F, 59S. (* 19F, 40S).
**Weight:** 57.00 tonnes.     **Maximum Speed:** 100 km/h.
**Length:** 23.400 m.

| | | | | |
|---|---|---|---|---|
| 515 520 EWAN | 515 529 EWAN | 515 554 EWAN | 515 591 EWAN | 515 616 (Z) |
| 515 522 EWAN | 515 536 EWAN | 515 556 EWAN | 515 604 EWAN | 515 633 (Z) |
| 515 525 EWAN | 515 548 EWAN | 515 557 (Z) | 515 605 (Z) | 515 636 (Z) |
| 515 526 EWAN | 515 549 (Z) | 515 566 (Z) | 515 608 (Z) | 515 643* EWAN |
| 515 528 (Z) | 515 551 EWAN | 515 580 EWAN | 515 611 EWAN | 515 645* EWAN |

## CLASS 815     Battery Unit Trailer

**Built:** 1956-64 by O & K/Rathgeber.
**Wheel Arrangement:** 2-2.     **Seats:** 81S.
**Weight:** 23.00 tonnes.     **Maximum Speed:** 100 km/h.
**Length:** 23.400 m.

| | | | | |
|---|---|---|---|---|
| 815 617 EWAN | 815 696 EWAN | 815 706 EWAN | 815 713 EWAN | 815 807 EWAN |
| 815 672 EWAN | 815 697 EWAN | 815 711 EWAN | | |

# DIESEL MULTIPLE UNITS

## CLASS 610     2-Car Unit

**Built:** 1992-93 by Duewag/MAN.
**Engine:** One MTU 12V183TD12 of 485 kW per car.
**Transmission:** Electric. ABB/AEG/Siemens.
**Wheel Arrangement:** 2-A1 + 1A-A1.     **Seats:** 68S + 16F, 46S.
**Weight:**     **Maximum Speed:** 160 km/h.
**Length:** 26.306 + 26.306 m.

| | | | | | | | | |
|---|---|---|---|---|---|---|---|---|
| 610 001 | 610 501 | NN1 | 610 008 | 610 508 | NN1 | 610 015 | 610 515 | NN1 |
| 610 002 | 610 502 | NN1 | 610 009 | 610 509 | NN1 | 610 016 | 610 516 | NN1 |
| 610 003 | 610 503 | NN1 | 610 010 | 610 510 | NN1 | 610 017 | 610 517 | NN1 |
| 610 004 | 610 504 | NN1 | 610 011 | 610 511 | NN1 | 610 018 | 610 518 | NN1 |
| 610 005 | 610 505 | NN1 | 610 012 | 610 512 | NN1 | 610 019 | 610 519 | NN1 |
| 610 006 | 610 506 | NN1 | 610 013 | 610 513 | NN1 | 610 020 | 610 520 | NN1 |
| 610 007 | 610 507 | NN1 | 610 014 | 610 514 | NN1 | | | |

## CLASS 611      2-Car Unit

**Built:** On order from AEG/DWA.
**Engine:**
**Transmission:** Hydraulic. Voith.
**Wheel Arrangement:**      **Seats:**
**Weight:**      **Maximum Speed:**
**Length:**

| | | | | | |
|---|---|---|---|---|---|
| 611 001 | 611 501 | 611 018 | 611 518 | 611 035 | 611 535 |
| 611 002 | 611 502 | 611 019 | 611 519 | 611 036 | 611 536 |
| 611 003 | 611 503 | 611 020 | 611 520 | 611 037 | 611 537 |
| 611 004 | 611 504 | 611 021 | 611 521 | 611 038 | 611 538 |
| 611 005 | 611 505 | 611 022 | 611 522 | 611 039 | 611 539 |
| 611 006 | 611 506 | 611 023 | 611 523 | 611 040 | 611 540 |
| 611 007 | 611 507 | 611 024 | 611 524 | 611 041 | 611 541 |
| 611 008 | 611 508 | 611 025 | 611 525 | 611 042 | 611 542 |
| 611 009 | 611 509 | 611 026 | 611 526 | 611 043 | 611 543 |
| 611 010 | 611 510 | 611 027 | 611 527 | 611 044 | 611 544 |
| 611 011 | 611 511 | 611 028 | 611 528 | 611 045 | 611 545 |
| 611 012 | 611 512 | 611 029 | 611 529 | 611 046 | 611 546 |
| 611 013 | 611 513 | 611 030 | 611 530 | 611 047 | 611 547 |
| 611 014 | 611 514 | 611 031 | 611 531 | 611 048 | 611 548 |
| 611 015 | 611 515 | 611 032 | 611 532 | 611 049 | 611 549 |
| 611 016 | 611 516 | 611 033 | 611 533 | 611 050 | 611 550 |
| 611 017 | 611 517 | 611 034 | 611 534 | | |

## CLASS 614      Power Car

**Built:** 1971-76 by O & K/Uerdingen.
**Engine:** One MAN of 331 kW.
**Transmission:** Hydraulic. Voith.
**Wheel Arrangement:** B-2.      **Seats:** 12F, 58S.
**Weight:** 46.00 tonnes.      **Maximum Speed:** 140 km/h.
**Length:** 26.650 m.
**Note:** Work with Class 914 and 934 trailers as 3 or 4-car sets.

| | | | | | | | | | | | |
|---|---|---|---|---|---|---|---|---|---|---|---|
| 614 001 | NN1 | 614 018 | NN1 | 614 035 | NN1 | 614 052 | HBS | 614 069 | HBS |
| 614 002 | NN1 | 614 019 | NN1 | 614 036 | NN1 | 614 053 | HBS | 614 070 | HBS |
| 614 003 | NN1 | 614 020 | NN1 | 614 037 | NN1 | 614 054 | NN1 | 614 071 | NN1 |
| 614 004 | NN1 | 614 021 | NN1 | 614 038 | NN1 | 614 055 | HBS | 614 072 | NN1 |
| 614 005 | NN1 | 614 022 | NN1 | 614 039 | NN1 | 614 056 | HBS | 614 073 | HBS |
| 614 006 | NN1 | 614 023 | NN1 | 614 040 | NN1 | 614 057 | HBS | 614 074 | HBS |
| 614 007 | NN1 | 614 024 | NN1 | 614 041 | NN1 | 614 058 | HBS | 614 075 | NN1 |
| 614 008 | NN1 | 614 025 | NN1 | 614 042 | NN1 | 614 059 | HBS | 614 076 | NN1 |
| 614 009 | NN1 | 614 026 | NN1 | 614 043 | NN1 | 614 060 | HBS | 614 077 | HBS |
| 614 010 | NN1 | 614 027 | NN1 | 614 044 | NN1 | 614 061 | HBS | 614 078 | HBS |
| 614 011 | NN1 | 614 028 | NN1 | 614 045 | NN1 | 614 062 | HBS | 614 079 | HBS |
| 614 012 | NN1 | 614 029 | NN1 | 614 046 | NN1 | 614 063 | HBS | 614 080 | HBS |
| 614 013 | NN1 | 614 030 | NN1 | 614 047 | HBS | 614 064 | HBS | 614 081 | HBS |
| 614 014 | NN1 | 614 031 | NN1 | 614 048 | HBS | 614 065 | HBS | 614 082 | HBS |
| 614 015 | NN1 | 614 032 | NN1 | 614 049 | HBS | 614 066 | HBS | 614 083 | NN1 |
| 614 016 | NN1 | 614 033 | NN1 | 614 050 | HBS | 614 067 | HBS | 614 084 | NN1 |
| 614 017 | NN1 | 614 034 | NN1 | 614 051 | NN1 | 614 068 | HBS | | |

# CLASS 914                    Trailer Second

**Built:** 1971-76 by O & K/Uerdingen.
**Wheel Arrangement:** 2-2.            **Seats:** 88S.
**Weight:** 46.00 tonnes.          **Maximum Speed:** 140 km/h.
**Length:** 26.160 m.
**Note:** Work with Classes 614 and 934 in 3 and 4-car sets.

| | | | | | | | | | |
|---|---|---|---|---|---|---|---|---|---|
| 914 001 | NN1 | 914 010 | NN1 | 914 019 | NN1 | 914 027 | HBS | 914 035 | NN1 |
| 914 002 | NN1 | 914 011 | NN1 | 914 020 | NN1 | 914 028 | HBS | 914 036 | HBS |
| 914 003 | NN1 | 914 012 | NN1 | 914 021 | NN1 | 914 029 | HBS | 914 037 | HBS |
| 914 004 | NN1 | 914 013 | NN1 | 914 022 | NN1 | 914 030 | HBS | 914 038 | NN1 |
| 914 005 | NN1 | 914 014 | NN1 | 914 023 | NN1 | 914 031 | HBS | 914 039 | HBS |
| 914 006 | NN1 | 914 015 | NN1 | 914 024 | HBS | 914 032 | NN1 | 914 040 | HBS |
| 914 007 | NN1 | 914 016 | NN1 | 914 025 | HBS | 914 033 | HBS | 914 041 | HBS |
| 914 008 | NN1 | 914 017 | NN1 | 914 026 | NN1 | 914 034 | HBS | 914 042 | HBS |
| 914 009 | NN1 | 914 018 | NN1 | | | | | | |

# CLASS 624                    Power Car

**Built:** 1961-66 by Uerdingen. Some refurbished 1991 onwards by DB, Kassel Works.
**Engine:** One MAN of 331 kW.
**Transmission:** Hydraulic. Voith.
**Wheel Arrangement:** B-2.        **Seats:** 12F, 58S (refurbished 12F, 66S).
**Weight:** 44.00 (refurbished 51.00) tonnes.    **Maximum Speed:** 120 km/h.
**Length:** 26.650 m.
**Note:** Work with Class 924 trailers in 3-car sets.

| | | | | | | | | | |
|---|---|---|---|---|---|---|---|---|---|
| 624 501 | HO | 624 606 | HO | 624 628 | HO | 624 641 | HO | 624 669 | HO |
| 624 502 | HO | 624 609 | HO | 624 629 | HO | 624 643 | HO | 624 670 | HO |
| 624 503 | HO | 624 611 | HO | 624 630 | HO | 624 644 | HO | 624 671 | HO |
| 624 504 | HO | 624 612 | HO | 624 631 | HO | 624 645 | HO | 624 672 | HO |
| 624 505 | HO | 624 615 | HO | 624 632 | HO | 624 646 | HO | 624 673 | HO |
| 624 506 | HO | 624 616 | HO | 624 634 | HO | 624 647 | HO | 624 674 | HO |
| 624 507 | HO | 624 620 | HO | 624 635 | HO | 624 648 | HO | 624 675 | HO |
| 624 508 | HO | 624 622 | HO | 624 636 | HO | 624 649 | HO | 624 676 | HO |
| 624 601 | HO | 624 623 | HO | 624 637 | HO | 624 650 | HO | 624 677 | HO |
| 624 602 | HO | 624 624 | HO | 624 638 | HO | 624 662 | HO | 624 678 | HO |
| 624 604 | HO | 624 625 | HO | 624 639 | HO | 624 667 | HO | 624 679 | HO |
| 624 605 | HO | 624 626 | HO | 624 640 | HO | 624 668 | HO | 624 680 | HO |

# CLASS 924.2                    Bar Car

**Built:** 1961-66 by Uerdingen.
**Wheel Arrangement:** 2-2.        **Seats:** 24S.
**Weight:** 32.00 tonnes.      **Maximum Speed:** 120 km/h.
**Length:** 26.160 m.
**Note:** Work as required with Classes 624 and 924.4 in 3 and 4-car units.

| | | | |
|---|---|---|---|
| 924 201 | HO | 924 202 | HO |

# CLASS 924.4                  Trailer Second

**Built:** 1961-66 by Uerdingen. Some refurbished 1991 onwards by DB, Kassel Works.
**Wheel Arrangement:** 2-2.        **Seats:** 88S.
**Weight:** 29.00-31.00 (refurbished 38.00-40.00) tonnes.
**Length:** 26.160 m.        **Maximum Speed:** 120 km/h.
**Note:** Work with Class 624 in 3-car units.

| | | | | | | | | | |
|---|---|---|---|---|---|---|---|---|---|
| 924 401 | HO | 924 409 | HO | 924 417 | HO | 924 427 | HO | 924 437 | HO |
| 924 402 | HO | 924 410 | HO | 924 419 | HO | 924 428 | HO | 924 438 | HO |
| 924 403 | HO | 924 411 | HO | 924 420 | HO | 924 429 | HO | 924 439 | HO |
| 924 404 | HO | 924 412 | HO | 924 421 | HO | 924 430 | HO | 924 440 | HO |
| 924 405 | HO | 924 413 | HO | 924 423 | HO | 924 431 | HO | 924 501 | HO |
| 924 406 | HO | 924 414 | HO | 924 424 | HO | 924 433 | HO | 924 502 | HO |
| 924 407 | HO | 924 415 | HO | 924 425 | HO | 924 434 | HO | 924 505 | HO |
| 924 408 | HO | 924 416 | HO | 924 426 | HO | 924 436 | HO | | |

# CLASS 627.0 Single Unit

**Built:** 1974-75 by LHB.
**Engine:** One MaK of 260 kW.
**Transmission:** Hydraulic. Voith.
**Wheel Arrangement:** B-2
**Weight:** 33.00 tonnes.
**Length:** 22.500 m.

**Seats:** 64S.
**Maximum Speed:** 120 km/h.

| | | | | |
|---|---|---|---|---|
| 627 001 TT | 627 003 TT | 627 005 TT | 627 007 TT | 627 008 TT |
| 627 002 TT | 627 004 TT | 627 006 TT | | |

# CLASS 627.1 Single Unit

**Built:** 1981-82 by LHB.
**Engine:** One MaK of 287 kW.
**Transmission:** Hydraulic. Voith.
**Wheel Arrangement:** B-2.
**Weight:** 36.00 tonnes.
**Length:** 22.500 m.

**Seats:** 80S.
**Maximum Speed:** 120 km/h.

| | | | | |
|---|---|---|---|---|
| 627 101 MKP | 627 102 MKP | 627 103 MKP | 627 104 MKP | 627 105 MKP |

# CLASS 628.0 2-Car Unit

**Built:** 1974-75 by LHB/Uerdingen.
**Engine:** One of 260 kW per power car.
**Transmission:** Hydraulic. Voith.
**Wheel Arrangement:** B-2 + 2-B.
**Weight:** 32.50 + 31.50 tonnes.
**Length:** 22.175 + 22.175 m.

**Seats:** 60S + 76S.
**Maximum Speed:** 120 km/h.

| | | | | | | | | |
|---|---|---|---|---|---|---|---|---|
| 628 001 | 628 011 | MKP | 628 004 | 628 014 | MKP | 628 009 | 628 019 | MKP |
| 628 002 | 628 012 | MKP | 628 005 | 628 015 | MKP | 628 010 | 628 020 | MKP |
| 628 003 | 628 013 | MKP | 628 008 | 628 018 | MKP | | | |

# CLASSES 628.0 & 928.0 2-Car Unit

**Built:** 1974-75 by LHB/Uerdingen. Trailers converted from power cars in 1991.
**Engine:** One of 260 kW per power car.
**Transmission:** Hydraulic. Voith.
**Wheel Arrangement:** B-2 + 2-2.
**Weight:** 32.50 + 31.50 tonnes.
**Length:** 22.175 + 22.175 m.

**Seats:** 60S + 76S.
**Maximum Speed:** 120 km/h.

| | | | | | | | | |
|---|---|---|---|---|---|---|---|---|
| 628 021 | 928 021 | MKP | 628 023 | 928 023 | MKP | 628 024 | 928 024 | MKP |
| 628 022 | 928 022 | MKP | | | | | | |

# CLASSES 628.1 & 928.1 2-Car Unit

**Built:** 1981-82 by LHB/Uerdingen.
**Engine:** One of 260 kW per power car.
**Transmission:** Hydraulic. Voith.
**Wheel Arrangement:** B-2 + 2-2.
**Weight:** 36.00 + 27.00 tonnes.
**Length:** 22.175 + 22.175 m.

**Seats:** 64S + 64S.
**Maximum Speed:** 120 km/h.

| | | | | | | | | |
|---|---|---|---|---|---|---|---|---|
| 628 101 | 928 101 | MKP | 628 102 | 928 102 | MKP | 628 103 | 928 103 | MKP |

# CLASSES 628.2 & 928.2 2-Car Unit

**Built:** 1987-89 by Duewag/LHB/MBB/Uerdingen.
**Engine:** One Daimler-Benz OM444A of 410 kW.
**Transmission:** Hydraulic. Voith T320rz.
**Wheel Arrangement:** B-2 + 2-2.
**Weight:** 40.00 + 28.00 tonnes.
**Length:** 22.700 + 22.700 m.

**Seats:** 64S + 10F, 48S.
**Maximum Speed:** 120 km/h.

# GERMANY

| | | | | | | | | |
|---|---|---|---|---|---|---|---|---|
| 628 201 | 928 201 | AK | 628 251 | 928 251 | RK | 628 301 | 928 301 | RK |
| 628 202 | 928 202 | AK | 628 252 | 928 252 | RK | 628 302 | 928 302 | RK |
| 628 203 | 928 203 | AK | 628 253 | 928 253 | RK | 628 303 | 928 303 | RK |
| 628 204 | 928 204 | AK | 628 254 | 928 254 | RK | 628 304 | 928 304 | RK |
| 628 205 | 928 205 | AK | 628 255 | 928 255 | RK | 628 305 | 928 305 | RK |
| 628 206 | 928 206 | AK | 628 256 | 928 256 | RK | 628 306 | 928 306 | RK |
| 628 207 | 928 207 | AK | 628 257 | 928 257 | RK | 628 307 | 928 307 | RK |
| 628 208 | 928 208 | AK | 628 258 | 928 258 | RK | 628 308 | 928 308 | RK |
| 628 209 | 928 209 | AK | 628 259 | 928 259 | RK | 628 309 | 928 309 | RK |
| 628 210 | 928 210 | AK | 628 260 | 928 260 | RK | 628 310 | 928 310 | RK |
| 628 211 | 928 211 | AK | 628 261 | 928 261 | RK | 628 311 | 928 311 | RK |
| 628 212 | 928 212 | AK | 628 262 | 928 262 | RK | 628 312 | 928 312 | RK |
| 628 213 | 928 213 | AK | 628 263 | 928 263 | RK | 628 313 | 928 313 | RK |
| 628 214 | 928 214 | AK | 628 264 | 928 264 | RK | 628 314 | 928 314 | RK |
| 628 215 | 928 215 | AK | 628 265 | 928 265 | RK | 628 315 | 928 315 | RK |
| 628 216 | 928 216 | AK | 628 266 | 928 266 | RK | 628 316 | 928 316 | RK |
| 628 217 | 928 217 | AK | 628 267 | 928 267 | RK | 628 317 | 928 317 | RK |
| 628 218 | 928 218 | AK | 628 268 | 928 268 | RK | 628 318 | 928 318 | RK |
| 628 219 | 928 219 | AK | 628 269 | 928 269 | RK | 628 319 | 928 319 | RK |
| 628 220 | 928 220 | AK | 628 270 | 928 270 | RK | 628 320 | 928 320 | RK |
| 628 221 | 928 221 | AK | 628 271 | 928 271 | RK | 628 321 | 928 321 | RK |
| 628 222 | 928 222 | AK | 628 272 | 928 272 | RK | 628 322 | 928 322 | RK |
| 628 223 | 928 223 | FG | 628 273 | 928 273 | RK | 628 323 | 928 323 | FG |
| 628 224 | 928 224 | FG | 628 274 | 928 274 | RK | 628 324 | 928 324 | FG |
| 628 225 | 928 225 | FG | 628 275 | 928 275 | RK | 628 325 | 928 325 | FG |
| 628 226 | 928 226 | FG | 628 276 | 928 276 | RK | 628 326 | 928 326 | FG |
| 628 227 | 928 227 | FG | 628 277 | 928 277 | RK | 628 327 | 928 327 | FG |
| 628 228 | 928 228 | FG | 628 278 | 928 278 | RK | 628 328 | 928 328 | FG |
| 628 229 | 928 229 | MKP | 628 279 | 928 279 | RK | 628 329 | 928 329 | FG |
| 628 230 | 928 230 | MKP | 628 280 | 928 280 | RK | 628 330 | 928 330 | FG |
| 628 231 | 928 231 | MKP | 628 281 | 928 281 | RK | 628 331 | 928 331 | FG |
| 628 232 | 928 232 | MKP | 628 282 | 928 282 | RK | 628 332 | 928 332 | FG |
| 628 233 | 928 233 | MKP | 628 283 | 928 283 | RK | 628 333 | 928 333 | FG |
| 628 234 | 928 234 | MKP | 628 284 | 928 284 | RK | 628 334 | 928 334 | FG |
| 628 235 | 928 235 | MKP | 628 285 | 928 285 | RK | 628 335 | 928 335 | FG |
| 628 236 | 928 236 | MKP | 628 286 | 928 286 | RK | 628 336 | 928 336 | MKP |
| 628 237 | 928 237 | MKP | 628 287 | 928 287 | RK | 628 337 | 928 337 | MKP |
| 628 238 | 928 238 | MKP | 628 288 | 928 288 | RK | 628 338 | 928 338 | MKP |
| 628 239 | 928 239 | MKP | 628 289 | 928 289 | RK | 628 339 | 928 339 | MKP |
| 628 240 | 928 240 | MKP | 628 290 | 928 290 | RK | 628 340 | 928 340 | MKP |
| 628 241 | 928 241 | MKP | 628 291 | 928 291 | RK | 628 341 | 928 341 | MKP |
| 628 242 | 928 242 | MKP | 628 292 | 928 292 | RK | 628 342 | 928 342 | MKP |
| 628 243 | 928 243 | MKP | 628 293 | 928 293 | RK | 628 343 | 928 343 | MKP |
| 628 244 | 928 244 | MKP | 628 294 | 928 294 | RK | 628 344 | 928 344 | MKP |
| 628 245 | 928 245 | MKP | 628 295 | 928 295 | RK | 628 345 | 928 345 | MKP |
| 628 246 | 928 246 | MKP | 628 296 | 928 296 | RK | 628 346 | 928 346 | MKP |
| 628 247 | 928 247 | MKP | 628 297 | 928 297 | RK | 628 347 | 928 347 | MKP |
| 628 248 | 928 248 | MKP | 628 298 | 928 298 | RK | 628 348 | 928 348 | MKP |
| 628 249 | 928 249 | MKP | 628 299 | 928 299 | RK | 628 349 | 928 349 | MKP |
| 628 250 | 928 250 | MKP | 628 300 | 928 300 | RK | 628 350 | 928 350 | MKP |

## CLASSES 628.4 & 928.4 — 2-Car Unit

**Built:** 1992-95 by Duewag/LHB/LEW.
**Engine:** One Daimler Benz of 485 kW per power car.
**Transmission:** Hydraulic. Voith.
**Wheel Arrangement:** B-2 + 2-2.
**Weight:**
**Length:** 23.260 m + 23.260 m.

**Seats:** 64S + 10F, 48S.
**Maximum Speed:** 140 km/h.

| | | | | | | | | |
|---|---|---|---|---|---|---|---|---|
| 628 401 | 928 401 | NHO | 628 405 | 928 405 | NHO | 628 409 | 928 409 | NHO |
| 628 402 | 928 402 | NHO | 628 406 | 928 406 | NHO | 628 410 | 928 410 | NHO |
| 628 403 | 928 403 | NHO | 628 407 | 928 407 | NHO | 628 411 | 928 411 | NHO |
| 628 404 | 928 404 | NHO | 628 408 | 928 408 | NHO | 628 412 | 928 412 | NHO |

| | | | | | | | | |
|---|---|---|---|---|---|---|---|---|
| 628 413 | 928 413 | NHO | 628 476 | 928 476 | SKL | 628 539 | 928 539 | HBS |
| 628 414 | 928 414 | NHO | 628 477 | 928 477 | SKL | 628 540 | 928 540 | HBS |
| 628 415 | 928 415 | NHO | 628 478 | 928 478 | SKL | 628 541 | 928 541 | HBS |
| 628 416 | 928 416 | NHO | 628 479 | 928 479 | SKL | 628 542 | 928 542 | HBS |
| 628 417 | 928 417 | NHO | 628 480 | 928 480 | SKL | 628 543 | 928 543 | HBS |
| 628 418 | 928 418 | NHO | 628 481 | 928 481 | SKL | 628 544 | 928 544 | HBS |
| 628 419 | 928 419 | NHO | 628 482 | 928 482 | SKL | 628 545 | 928 545 | HBS |
| 628 420 | 928 420 | NHO | 628 483 | 928 483 | SKL | 628 546 | 928 546 | HBS |
| 628 421 | 928 421 | NHO | 628 484 | 928 484 | SKL | 628 547 | 928 547 | HBS |
| 628 422 | 928 422 | NHO | 628 485 | 928 485 | SKL | 628 548 | 928 548 | HBS |
| 628 423 | 928 423 | NHO | 628 486 | 928 486 | SKL | 628 549 | 928 549 | HBS |
| 628 424 | 928 424 | NHO | 628 487 | 928 487 | SKL | 628 550 | 928 550 | HBS |
| 628 425 | 928 425 | NHO | 628 488 | 928 488 | SKL | 628 551 | 928 551 | HBS |
| 628 426 | 928 426 | NHO | 628 489 | 928 489 | SKL | 628 552 | 928 552 | HBS |
| 628 427 | 928 427 | NHO | 628 490 | 928 490 | SKL | 628 553 | 928 553 | HBS |
| 628 428 | 928 428 | FG | 628 491 | 928 491 | SKL | 628 554 | 928 554 | HBS |
| 628 429 | 928 429 | FG | 628 492 | 928 492 | SKL | 628 555 | 928 555 | HBS |
| 628 430 | 928 430 | FG | 628 493 | 928 493 | SKL | 628 556 | 928 556 | HBS |
| 628 431 | 928 431 | FG | 628 494 | 928 494 | FG | 628 557 | 928 557 | HBS |
| 628 432 | 928 432 | FG | 628 495 | 928 495 | EE | 628 558 | 928 558 | MMF |
| 628 433 | 928 433 | FG | 628 496 | 928 496 | EE | 628 559 | 928 559 | MMF |
| 628 434 | 928 434 | FG | 628 497 | 928 497 | EE | 628 560 | 928 560 | MMF |
| 628 435 | 928 435 | FG | 628 498 | 928 498 | EE | 628 561 | 928 561 | MMF |
| 628 436 | 928 436 | FG | 628 499 | 928 499 | EE | 628 562 | 928 562 | MMF |
| 628 437 | 928 437 | FG | 628 500 | 928 500 | EE | 628 563 | 928 563 | MMF |
| 628 438 | 928 438 | FG | 628 501 | 928 501 | EE | 628 564 | 928 564 | MMF |
| 628 439 | 928 439 | FG | 628 502 | 928 502 | EE | 628 565 | 928 565 | MMF |
| 628 440 | 928 440 | FG | 628 503 | 928 503 | EE | 628 566 | 928 566 | MMF |
| 628 441 | 928 441 | FG | 628 504 | 928 504 | EE | 628 567 | 928 567 | MMF |
| 628 442 | 928 442 | FG | 628 505 | 928 505 | SKL | 628 568 | 928 568 | MMF |
| 628 443 | 928 443 | FG | 628 506 | 928 506 | SKL | 628 569 | 928 569 | MMF |
| 628 444 | 928 444 | FG | 628 507 | 928 507 | EE | 628 570 | 928 570 | MMF |
| 628 445 | 928 445 | FG | 628 508 | 928 508 | EE | 628 571 | 928 571 | MMF |
| 628 446 | 928 446 | FG | 628 509 | 928 509 | EE | 628 572 | 928 572 | MMF |
| 628 447 | 928 447 | FG | 628 510 | 928 510 | EE | 628 573 | 928 573 | MMF |
| 628 448 | 928 448 | FG | 628 511 | 928 511 | EE | 628 574 | 928 574 | MMF |
| 628 449 | 928 449 | FG | 628 512 | 928 512 | EE | 628 575 | 928 575 | MMF |
| 628 450 | 928 450 | FG | 628 513 | 928 513 | EE | 628 576 | 928 576 | MMF |
| 628 451 | 928 451 | FG | 628 514 | 928 514 | EE | 628 577 | 928 577 | MMF |
| 628 452 | 928 452 | FG | 628 515 | 928 515 | EE | 628 578 | 928 578 | MMF |
| 628 453 | 928 453 | FG | 628 516 | 928 516 | EE | 628 579 | 928 579 | MMF |
| 628 454 | 928 454 | FG | 628 517 | 928 517 | EE | 628 580 | 928 580 | |
| 628 455 | 928 455 | FG | 628 518 | 928 518 | EE | 628 581 | 928 581 | LL1 |
| 628 456 | 928 456 | FG | 628 519 | 928 519 | EE | 628 582 | 928 582 | LL1 |
| 628 457 | 928 457 | FG | 628 520 | 928 520 | EE | 628 583 | 928 583 | AK |
| 628 458 | 928 458 | FG | 628 521 | 928 521 | EE | 628 584 | 928 584 | AK |
| 628 459 | 928 459 | FG | 628 522 | 928 522 | SKL | 628 585 | 928 585 | LL1 |
| 628 460 | 928 460 | FG | 628 523 | 928 523 | SKL | 628 586 | 928 586 | LL1 |
| 628 461 | 928 461 | FG | 628 524 | 928 524 | SKL | 628 587 | 928 587 | LL1 |
| 628 462 | 928 462 | SKL | 628 525 | 928 525 | EE | 628 588 | 928 588 | LL1 |
| 628 463 | 928 463 | SKL | 628 526 | 928 526 | EE | 628 589 | 928 589 | LL1 |
| 628 464 | 928 464 | RK | 628 527 | 928 527 | HBS | 628 590 | 928 590 | LL1 |
| 628 465 | 928 465 | SKL | 628 528 | 928 528 | HBS | 628 591 | 928 591 | LL1 |
| 628 466 | 928 466 | SKL | 628 529 | 928 529 | HBS | 628 592 | 928 592 | LL1 |
| 628 467 | 928 467 | SKL | 628 530 | 928 530 | HBS | 628 593 | 928 593 | LL1 |
| 628 468 | 928 468 | SKL | 628 531 | 928 531 | HBS | 628 594 | 928 594 | LL1 |
| 628 469 | 928 469 | SKL | 628 532 | 928 532 | HBS | 628 595 | 928 595 | LL1 |
| 628 470 | 928 470 | SKL | 628 533 | 928 533 | HBS | 628 596 | 928 596 | LL1 |
| 628 471 | 928 471 | SKL | 628 534 | 928 534 | HBS | 628 597 | 928 597 | LL1 |
| 628 472 | 928 472 | SKL | 628 535 | 928 535 | HBS | 628 598 | 928 598 | LL1 |
| 628 473 | 928 473 | SKL | 628 536 | 928 536 | HBS | 628 599 | 928 599 | LL1 |
| 628 474 | 928 474 | SKL | 628 537 | 928 537 | HBS | 628 600 | 928 600 | LL1 |
| 628 475 | 928 475 | SKL | 628 538 | 928 538 | HBS | 628 601 | 928 601 | LL1 |

# GERMANY

| 628 602 | 928 602 | LL1 | 628 637 | 928 637 | WNT | 628 671 | 928 671 | FG |
|---|---|---|---|---|---|---|---|---|
| 628 603 | 928 603 | LL1 | 628 638 | 928 638 | WNT | 628 672 | 928 672 | FG |
| 628 604 | 928 604 | LL1 | 628 639 | 928 639 | WNT | 628 673 | 928 673 | FG |
| 628 605 | 928 605 | LL1 | 628 640 | 928 640 | WNT | 628 674 | 928 674 | FG |
| 628 606 | 928 606 | LL1 | 628 641 | 928 641 | WNT | 628 675 | 928 675 | FG |
| 628 607 | 928 607 | LL1 | 628 642 | 928 642 | WNT | 628 676 | 928 676 | FG |
| 628 608 | 928 608 | LL1 | 628 643 | 928 643 | WNT | 628 677 | 928 677 | FG |
| 628 609 | 928 609 | LL1 | 628 644 | 928 644 | WNT | 628 678 | 928 678 | AK |
| 628 610 | 928 610 | LL1 | 628 645 | 928 645 | WNT | 628 679 | 928 679 | |
| 628 611 | 928 611 | LL1 | 628 646 | 928 646 | WNT | 628 680 | 928 680 | |
| 628 612 | 928 612 | LL1 | 628 647 | 928 647 | WNT | 628 681 | 928 681 | AK |
| 628 613 | 928 613 | LL1 | 628 648 | 928 648 | WNT | 628 682 | 928 682 | AK |
| 628 614 | 928 614 | HBS | 628 649 | 928 649 | WNT | 628 683 | 928 683 | AK |
| 628 615 | 928 615 | HBS | 628 650 | 928 650 | WNT | 628 684 | 928 684 | WNT |
| 628 616 | 928 616 | HBS | 628 651 | 928 651 | | 628 685 | 928 685 | WNT |
| 628 617 | 928 617 | HBS | 628 652 | 928 652 | | 628 686 | 928 686 | WNT |
| 628 618 | 928 618 | HBS | 628 653 | 928 653 | | 628 687 | 928 687 | WNT |
| 628 619 | 928 619 | HBS | 628 654 | 928 654 | | 628 688 | 928 688 | WNT |
| 628 620 | 928 620 | HBS | 628 655 | 928 655 | | 628 689 | 928 689 | FG |
| 628 621 | 928 621 | HBS | 628 656 | 928 656 | | 628 690 | 928 690 | WNT |
| 628 622 | 928 622 | | 628 657 | 928 657 | | 628 691 | 928 691 | WNT |
| 628 623 | 928 623 | | 628 658 | 928 658 | | 628 692 | 928 692 | MMF |
| 628 624 | 928 624 | | 628 659 | 928 659 | | 628 693 | 928 693 | FG |
| 628 625 | 928 625 | | 628 660 | 928 660 | | 628 694 | 928 694 | FG |
| 628 626 | 928 626 | | 628 661 | 928 661 | | 628 695 | 928 695 | WNT |
| 628 627 | 928 627 | | 628 662 | 928 662 | | 628 696 | 928 696 | |
| 628 628 | 928 628 | | 628 663 | 928 663 | | 628 697 | 928 697 | |
| 628 629 | 928 629 | | 628 664 | 928 664 | | 628 698 | 928 698 | |
| 628 630 | 928 630 | MMF | 628 665 | 928 665 | | 628 699 | 928 699 | |
| 628 631 | 928 631 | WNT | 628 666 | 928 666 | | 628 700 | 928 700 | |
| 628 632 | 928 632 | EE | 628 667 | 928 667 | | 628 701 | 928 701 | |
| 628 633 | 928 633 | WNT | 628 668 | 928 668 | | 628 702 | 928 702 | |
| 628 634 | 928 634 | WNT | 628 669 | 928 669 | | 628 703 | 928 703 | |
| 628 635 | 928 635 | WNT | 628 670 | 928 670 | FG | 628 704 | 928 704 | |
| 628 636 | 928 636 | WNT | | | | | | |

## CLASSES 628.9 & 629     2-Car Unit

**Built:** 1995 by Duewag.
**Engine:** One Daimler Benz of 485 kW per power car.
**Transmission:** Hydraulic. Voith.
**Wheel Arrangement:** B-2 + 2-B.　　　　**Seats:** 64S + 10F, 48S.
**Weight:**　　　　**Maximum Speed:** 140 km/h.
**Length:** 23.260 m + 23.260 m.

| 628 901 | 629 001 | SKL | 628 903 | 629 003 | SKL | 628 905 | 629 005 | SKL |
|---|---|---|---|---|---|---|---|---|
| 628 902 | 629 002 | SKL | 628 904 | 629 004 | SKL | | | |

## CLASS 634     Power Car

**Built:** 1964-66 by Uerdingen. Some refurbished 1991 onwards by DB, Kassel Works.
**Engine:** One MAN of 331 kW.
**Transmission:** Hydraulic. Voith.
**Wheel Arrangement:** B-2.　　　　**Seats:** 12F, 58S (refurbished 12F, 66S).
**Weight:** 44.00 (refurbished 51.00) tonnes.　　　　**Maximum Speed:** 120 km/h.
**Length:** 26.650 m.
**Note:** Work with Class 934 trailers in 3-car sets.

| 634 603 | HO | 634 614 | HO | 634 627 | HO | 634 654 | HO | 634 659 | HO |
|---|---|---|---|---|---|---|---|---|---|
| 634 607 | HO | 634 617 | HO | 634 633 | HO | 634 655 | HO | 634 661 | HO |
| 634 608 | HO | 634 618 | HO | 634 651 | HO | 634 656 | HO | 634 664 | HO |
| 634 610 | HO | 634 619 | HO | 634 652 | HO | 634 657 | HO | 634 665 | HO |
| 634 613 | HO | 634 621 | HO | 634 653 | HO | 634 658 | HO | 634 666 | HO |

# CLASS 934          Trailer Second

**Built:** 1961-68 by Uerdingen. Some refurbished 1991 onwards by DB, Kassel Works.
**Wheel Arrangement:** 2-2.      **Seats:** 88S († 24S).
**Weight:** 31.00 (* 29.00; † 32.00) tonnes.      **Maximum Speed:** 120 (NN1 allocation 140) km/h.
**Length:** 26.160 m.
**Notes:** Work with Classes 614, 634 and 914 in 3 and 4-car units.

| | | | | | | | | | |
|---|---|---|---|---|---|---|---|---|---|
| 934 202† | HBS | 934 443 | HO | 934 447 | HO | 934 450 | HO | 934 453 | HO |
| 934 422 | HO | 934 444 | HO | 934 448 | HO | 934 451 | HO | 934 454 | HO |
| 934 441 | NN1 | 934 445 | HO | 934 449 | HO | 934 452 | HO | 934 506* | HO |
| 934 442 | NN1 | 934 446 | HO | | | | | | |

# CLASS 685          Power Car

**Built:** 1935 by Westwaggon.
**Engine:** One Maybach of 309 kW.
**Transmission:** Electric. AEG.
**Wheel Arrangement:** B-2.      **Seats:** 16F, 45S.
**Weight:** 46.10 tonnes.      **Maximum Speed:** 110 km/h.
**Length:** 21.870 m.
**Notes:** Previously DR Class 185. Works with Class 947 trailer.

685 254 WSR

# CLASS 947          Trailer

**Built:**
**Wheel Arrangement:** 2-2.      **Seats:**
**Weight:**      **Maximum Speed:** 110 km/h.
**Length:**
**Notes:** Previously DR Class 185. Works with Class 685 power car.
947 052 WSR

## DEPARTMENTAL VEHICLES

# CLASS 701          OHL Maintenance Car

**Built:** 1954-64 by MBB/Rathgeber.
**Engines:** Two Büssing of 110 kW each.
**Transmission:** Mechanical.
**Wheel Arrangement:** A-A.      **Length:** 13.950 m.
**Weight:** 24.60 tonnes.      **Maximum Speed:** 90 km/h.

| | | | | | | | | | |
|---|---|---|---|---|---|---|---|---|---|
| 701 001 | HE | 701 019 | KK | 701 037 | MA | 701 063 | FK | 701 081 | HE |
| 701 002 | ESIE | 701 020 | FD | 701 038 | EDO | 701 064 | MIH | 701 082 | HB |
| 701 003 | EHM | 701 021 | RM | 701 039 | FF2 | 701 065 | MMF | 701 083 | HH |
| 701 004 | RM | 701 022 | NN1 | 701 040 | KK | 701 066 | NN1 | 701 084 | KM |
| 701 005 | NN1 | 701 023 | HG | 701 043 | NN1 | 701 067 | NN1 | 701 085 | KM |
| 701 006 | RK | 701 024 | FK | 701 044 | KK | 701 068 | SSH | 701 086 | RM |
| 701 007 | HG | 701 025 | TK | 701 045 | FK | 701 069 | TK | 701 087 | KK |
| 701 008 | FK | 701 026 | EOB | 701 046 | HO | 701 070 | TK | 701 088 | HO |
| 701 009 | HH | 701 027 | MH1 | 701 047 | FK | 701 071 | RO | 701 089 | NN1 |
| 701 010 | TK | 701 028 | MMF | 701 048 | EHG | 701 072 | FK | 701 090 | HH |
| 701 011 | RSI | 701 029 | NWH | 701 051 | ESIE | 701 073 | EOB | 701 091 | SSH |
| 701 012 | RK | 701 030 | NRH | 701 052 | EOB | 701 074 | FD | 701 095 | AH4 |
| 701 013 | MA | 701 031 | NRH | 701 053 | HH | 701 075 | HG | 701 096 | AH4 |
| 701 014 | FG | 701 032 | TH | 701 054 | HG | 701 076 | HH | 701 097 | AH4 |
| 701 015 | FF2 | 701 033 | TK | 701 058 | KM | 701 077 | HB | 701 098 | FK |
| 701 016 | RM | 701 034 | HG | 701 059 | SKL | 701 078 | HBS | 701 099 | FK |
| 701 017 | EHM | 701 035 | FD | 701 060 | KK | 701 079 | HB | 701 100 | FK |
| 701 018 | KK | 701 036 | KK | 701 062 | MA | 701 080 | HH | 701 101 | HBS |

# GERMANY

| | | | | | | | | | |
|---|---|---|---|---|---|---|---|---|---|
| 701 102 | FF2 | 701 113 | MH1 | 701 127 | HB | 701 146 | HH | 701 159 | MH1 |
| 701 103 | FF2 | 701 114 | EHM | 701 128 | RSI | 701 147 | STR | 701 160 | KK |
| 701 104 | FG | 701 116 | EOB | 701 130 | AH4 | 701 151 | TH | 701 161 | HG |
| 701 105 | FD | 701 117 | MIH | 701 139 | EHM | 701 152 | TH | 701 162 | HB |
| 701 106 | KK | 701 118 | HO | 701 140 | EDO | 701 153 | TK | 701 165 | MMF |
| 701 107 | KK | 701 119 | HH | 701 141 | EOB | 701 154 | NWH | 701 166 | TK |
| 701 108 | KK | 701 120 | EHM | 701 142 | FF2 | 701 155 | STR | 701 167 | EDO |
| 701 109 | RM | 701 121 | HO | 701 143 | FD | 701 156 | RM | 701 168 | RO |
| 701 110 | EOB | 701 122 | TK | 701 144 | MA | 701 157 | EHM | 701 169 | EHM |
| 701 111 | TK | 701 126 | HO | 701 145 | EDO | 701 158 | MH1 | 701 170 | HBS |
| 701 112 | EDO | | | | | | | | |

## CLASS 702                                    OHL Maintenance Car

**Built:** 1954-64 by MBB/Rathgeber as class 701. One engine removed 1967.
**Engine:** Büssing of 110 kW.
**Transmission:** Mechanical.
**Wheel Arrangement:** 1-A.          **Length:** 13.950 m.
**Weight:** 21.80 tonnes.            **Maximum Speed:** 90 km/h.

| | | | | | | | | | |
|---|---|---|---|---|---|---|---|---|---|
| 702 042 | EHG | 702 115 | MH1 | 702 131 | RHL | 702 136 | MH1 | 702 149 | TK |
| 702 049 | KK | 702 123 | MH1 | 702 132 | TK | 702 137 | MA | 702 150 | NN1 |
| 702 050 | KK | 702 124 | MH1 | 702 133 | TK | 702 138 | NRH | 702 163 | SSH |
| 702 055 | RF | 702 125 | MH1 | 702 134 | KK | 702 148 | RK | 702 164 | AH4 |
| 702 056 | RF | 702 129 | NN1 | 702 135 | HG | | | | |

## CLASS 704                                    OHL Maintenance Car

**Built:** 1977-78 by LHB.
**Engine:** MaK of 260 kW.
**Transmission:** Hydraulic.        **Length:** 22.500 m.
**Weight:**                         **Maximum Speed:** 120 km/h.

| | | | | | | | | | |
|---|---|---|---|---|---|---|---|---|---|
| 704 001 | HG | 704 002 | RM | 704 003 | AH4 | 704 004 | NWH | 704 005 | FK |

## CLASS 705                                  Tunnel Inspection Unit

**Built:** 1992 by Deutsche Plasser Baumaschinen.
**Engine:** of 367 kW.
**Transmission:**     **Length:**
**Weight:**                         **Maximum Speed:** 120 km/h.

705 001 RK

## CLASS 708.0                                  OHL Maintenance Car

**Built:** 1956-58 by Görlitz.
**Engine:** IFA of 99 kW (* Johannisthal of 111 kW).
**Transmission:** Mechanical. Strömungsmachinen.
**Wheel Arrangement:** A-1.          **Length:** 13.100 m.
**Weight:** 26.00 tonnes.            **Maximum Speed:** 70 km/h.
**Note:** Previously DR Class 188.0.

| | | | | | |
|---|---|---|---|---|---|
| 708 001 | BSE | 708 002 | (Z) | 708 006* | LL1 |

## CLASS 708.2                                  OHL Maintenance Car

**Built:** 1968-69 by Görlitz.
**Engine:** Johannisthal of 132 kW.
**Transmission:** Mechanical. Strömungsmachinen.
**Wheel Arrangement:** 1A-2.          **Length:** 19.300 m.
**Weight:** 43.00 tonnes.             **Maximum Speed:** 80 km/h.
**Note:** Previously DR Class 188.2.

| | | | | | | | | | |
|---|---|---|---|---|---|---|---|---|---|
| 708 200 | WR | 708 202 | DH | 708 203 | DC | 708 204 | BCS | 708 205 | LL1 |
| 708 201 | UN | | | | | | | | |

## CLASS 708.3           OHL Maintenance Car

**Built:** 1987-91 by Görlitz.
**Engine:** Johannisthal 6VD18/15 AL-2 of 331 kW.
**Transmission:** Mechanical. Strömungsmachinen.
**Wheel Arrangement:** B-2.       **Length:** 22.400 m.
**Weight:** 58.00 tonnes.      **Maximum Speed:** 100 km/h.
**Note:** Previously DR Class 188.3

| 708 301 | LS | 708 309 | LL1 | 708 317 | BSE | 708 324 | LS | 708 331 | LL1 |
|---|---|---|---|---|---|---|---|---|---|
| 708 302 | LL1 | 708 310 | LS | 708 318 | UEI | 708 325 | LL1 | 708 332 | WNT |
| 708 303 | DZW | 708 311 | WNT | 708 319 | BCS | 708 326 | WSR | 708 333 | UN |
| 708 304 | BSE | 708 312 | BCS | 708 320 | LL1 | 708 327 | WNT | 708 334 | BSE |
| 708 305 | BSE | 708 313 | WSR | 708 321 | DC | 708 328 | BHW | 708 335 | DH |
| 708 306 | BHW | 708 314 | WNT | 708 322 | BFG | 708 329 | DR | 708 336 | WNT |
| 708 307 | WNT | 708 315 | BSE | 708 323 | WP | 708 330 | LS | 708 337 | UE |
| 708 308 | WP | 708 316 | LL1 | | | | | | |

## CLASS 710         B-B

**Built:** 1964-78 by LEW.
**Engine:** Johannisthal 12 KVD 21 AL3 of 736 kW at 1500 rpm.
**Transmission:** Hydraulic. Strömungsmachinen GSR 30/5.7 AQ.
**Weight:** 62.00 tonnes.      **Maximum Speed:** 100 km/h.
**Length:** 14.240 m.      **Train Supply:** Not equipped.
**Note:** Previously DR Class 110.9.

| 710 961 | UEI | 710 963 | LM | 710 965 | LM | 710 967 | LL1 | 710 969 | WG |
|---|---|---|---|---|---|---|---|---|---|
| 710 962 | BPKR | 710 964 | BPKR | 710 966 | WP | 710 968 | WR | 710 970 | DC |

## CLASS 712       Tunnel Inspection Unit

**Built:** 1993 by DB, Kassel Works.
**Engine:**
**Transmission:**
**Wheel Arrangement:**    **Length:**
**Weight:**    **Maximum Speed:**

712 002 RK

## CLASS 714         B-B

**Built:** 1962-66 by MaK/Henschel/Jung/Deutz/Krauss-Maffei/Krupp/Esslingen as Class 212. Rebuilt as Tunnel Emergency Train locomotives in 1989-91 by MaK/DB Bremen Works.
**Engine:** Mercedes 12 V 652 TZ of 993 kW at 1500 rpm.
**Transmission:** Hydraulic. Voith L216rs.    **Length:** 12.300 m.
**Weight:** 63.00 tonnes.    **Maximum Speed:** 100 km/h.

| 714 033 | TK | 714 236 | HBS | 714 246 | NWH | 714 260 | RK | 714 277 | RK |
|---|---|---|---|---|---|---|---|---|---|
| 714 046 | TK | 714 244 | NWH | 714 251 | NWH | 714 271 | NWH | 714 352 | HBS |
| 714 235 | NWH | 714 245 | NWH | 714 257 | NWH | | | | |

## CLASS 716       Rotary Snow Plough

**Built:** 1994 by Beilhack.
**Engine:**
**Transmission:**    **Length:**
**Weight:**    **Maximum Speed:**

716 001 FFU    716 002 HBS

## CLASSES 719 & 720       Rail Testing Unit

**Built:** 1974.
**Engine:**
**Transmission:**    **Length:**
**Weight:**    **Maximum Speed:**

# GERMANY

719 001   720 001   719 501   HBS

## CLASS 723 — Radio Test Unit

**Built:** 1934 by Wismar.
**Engine:** One Maybach of 302 kW.
**Transmission:** Electric. SSW.
**Wheel Arrangement:** 2-Bo.
**Weight:** 42.00 tonnes.
**Length:** 20.930 m.
**Maximum Speed:** 100 km/h.

723 101 BBH

## CLASS 724 — Indusi Test Cars

**Built:** 1952-54 by Uerdingen/Rathgeber/Lüttgens/Orion/WMD.
**Engine:** Büssing of 110 kW.
**Transmission:** Mechanical.
**Wheel Arrangement:** A-1.
**Weight:** 13.30 tonnes.
**Length:** 13.265 m (724 003); 13.298 m (724 002).
**Maximum Speed:** 90 km/h.

724 002 KK        724 003  KK

## CLASSES 725/726 — Track Testing Unit

**Built:** 1959-62 by Uerdingen/MAN/WMD(Class 725); 1974 (Class 726).
**Engine:** Two Büssing of 110 kW each (Class 725 only).
**Transmission:** Mechanical.
**Wheel Arrangement:** Bo + 1-1.
**Weight:**
**Length:** 13.950 m.
**Maximum Speed:** 90 km/h.

725 001   726 001   MA      725 003   726 003   HBS      725 005   726 005   NN1
725 002   726 002   FD      725 004   726 004   KK

## CLASS 727 — Cab Signalling Test Car

**Built:** 1952 by Uerdingen.
**Engine:** Büssing of 110 kW.
**Transmission:** Mechanical.
**Wheel Arrangement:** A-1.
**Weight:**
**Length:** 13.265 m.
**Maximum Speed:** 90 km/h.

727 001 MH1

## CLASS 728 — Indusi Test Car

**Built:** 1962 by Uerdingen/MAN/WMD. Converted 1985.
**Engines:** Two Büssing of 110 kW each.
**Transmission:** Mechanical.
**Wheel Arrangement:** B.
**Weight:**
**Length:** 13.950 m.
**Maximum Speed:** 90 km/h.

728 001 KK

## CLASS 732 — De-Icing Unit

**Built:** 1959-63 by DWM/MAN/O & K/Rathgeber. Converted 1991.
**Batteries:** 520 kWh.
**Electrical Equipment:** AFA/DWM/Schaltbau/Siemens.
**Continuous Rating:** 200 kW.
**Wheel Arrangement:** Bo-2.
**Weight:**
**Length:** 23.400 m.
**Maximum Speed:** 100 km/h.

732 001 AOP

# CLASS 740     S & T Test Car

**Built:** 1959-62 by Uerdingen/MAN/WMD. Converted 1990-92.
**Engine:** Two Büssing of 110 kW each.
**Transmission:** Mechanical.
**Wheel Arrangement:** B.     **Length:** 13.950 m.
**Weight:**     **Maximum Speed:** 90 km/h.

| | | | | |
|---|---|---|---|---|
| 740 001 FK | 740 003 TK | 740 004 FK | 740 005 RK | 740 006 HG |
| 740 002 FK | | | | |

# 750 001     Co-Co

**Built:** 1965 by Henschel.
**Electrical Equipment:** Siemens.
**Continuous Rating:** 5950 kW.
**Weight:** 110.00 tonnes.     **Maximum Speed:** 200 km/h.
**Length:** 19.500 m.     **Train Supply:** Electric.

750 001 AH1

# 750 003     Co-Co

**Built:** 1970-74 by Henschel/Krauss-Maffei/Krupp.
**Electrical Equipment:** Siemens/AEG/BBC.
**Continuous Rating:** 7440 kW.
**Weight:** 114.00 tonnes.     **Maximum Speed:** 200 km/h.
**Length:** 20.200 m.     **Train Supply:** Electric.

750 003 AH1

# CLASS 751     Bo-Bo

**Built:** 1965 by Henschel/Krauss-Maffei/Krupp.
**Electrical Equipment:** BBC/Siemens/AEG.
**Continuous Rating:** 3620 kW.
**Weight:** 84.60 tonnes.     **Maximum Speed:** 150 km/h.
**Length:** 16.490 m.     **Train Supply:** Electric.

751 001 AH1

# CLASS 752     Bo-Bo

**Built:** 1979 by Henschel/Krauss-Maffei/Krupp.
**Electrical Equipment:** BBC.
**Continuous Rating:** 5600 kW.
**Weight:** 83.20 tonnes.     **Maximum Speed:** 200 km/h.
**Length:** 19.200 m.     **Train Supply:** Electric.

752 001 NN1    *Nürnberg*

# CLASS 753     B-B

**Built:** 1965 by Krupp.
**Engine:** Mercedes 16 V 652 TB of 1400 kW at 1500 rpm.
**Transmission:** Hydraulic. Voith L820brs.
**Weight:**     **Maximum Speed:** 140 km/h.
**Length:** 16.400 m.     **Train Supply:** Not equipped.

753 001 NRH    753 002 NRH

# CLASS 754     Co-Co

**Built:** 1970-72 by Voroshilovgrad.
**Engine:** Kolomna 5 D 49 of 2200 kW at 1000 rpm.
**Transmission:** Electric. Charkov ED118 traction motors.
**Weight:** 115.00 tonnes.     **Maximum Speed:** 140 km/h.

# GERMANY

**Length:** 20.620 m.  **Train Supply:** Electric.

754 101 LH1    754 102 (Z)

## CLASS 755 Bo-Bo

**Built:** 1991 by LEW.
**Electrical Equipment:** LEW.
**Continuous Rating:** 3500 kW.
**Weight:** 82.50 tonnes.  **Maximum Speed:** 140 km/h.
**Length:** 16.640 m.  **Train Supply:** Electric.

755 025 LH2

## CLASS 756 B

**Built:** 1960-63 by LKM. Rebuilt 1975-79 by DR Halle Works.
**Engine:** Johannisthal 6 KVD 18/15-1 SRW of 162 kW at 1510 rpm.
**Transmission:** Hydraulic. Strömungsmachinen GSU 20/4.2.
**Weight:** 21.50 tonnes.  **Length:** 6.940 m.
**Maximum Speed:** 42 km/h.

756 624 LH2

# DIESEL RAILBUSES

## CLASS 771 Railbus

**Built:** 1958-64 by Bautzen.
**Engine:** Roßlau of 132 kW or MAN of ??? kW.
**Transmission:** Mechanical. Strömungsmachinen.
**Wheel Arrangement:** 1-A.
**Weight:** 19.30 tonnes.  **Maximum Speed:** 90 km/h.
**Length:** 13.550 m.  **Seats:** 54S.
**Notes:** Previously DR Class 171.0. Operate with Class 971 trailers.

| | | | | | | | | | |
|---|---|---|---|---|---|---|---|---|---|
| 771 003 | DH | 771 016 | LL1 | 771 030 | WSR | 771 046 | WSR | 771 059 | WNT |
| 771 004 | LS | 771 017 | WSR | 771 031 | LG | 771 047 | UN | 771 060 | LS |
| 771 005 | WSR | 771 018 | UN | 771 032 | BCS | 771 048 | LHB | 771 061 | WSR |
| 771 006 | WSR | 772 319 | LM | 771 034 | LS | 771 049 | UN | 771 062 | DH |
| 771 007 | WSR | 771 020 | LG | 771 035 | UN | 771 050 | WSR | 771 063 | UN |
| 771 008 | UN | 771 021 | DH | 771 038 | LG | 771 051 | LS | 771 064 | BFG |
| 771 009 | LM | 771 022 | BFG | 771 039 | LM | 771 052 | WSR | 771 065 | WSR |
| 771 010 | LM | 771 023 | WSR | 771 040 | LHB | 771 054 | LM | 771 066 | LM |
| 771 011 | UN | 771 024 | LM | 771 041 | UN | 771 055 | UN | 771 067 | DH |
| 771 012 | LM | 771 026 | WSR | 771 042 | WP | 771 056 | LM | 771 068 | LS |
| 771 013 | WSR | 771 027 | LM | 771 043 | WW | 771 057 | UN | 771 069 | UN |
| 771 014 | WSR | 771 028 | UN | 771 044 | UM | 771 058 | WSR | 771 070 | WSR |
| 771 015 | UN | 771 029 | WSR | 771 045 | LM | | | | |

## CLASS 971.0 Railbus Non-Driving Trailer

**Built:** 1958-64 by Bautzen.
**Wheel Arrangement:** 1-1.  **Seats:** 54S.
**Weight:**   **Maximum Speed:** 90 km/h.
**Length:** 13.550 m.
**Note:** Previously DR Class 171.8.

| | | | | | | | | | |
|---|---|---|---|---|---|---|---|---|---|
| 971 003 | LS | 971 011 | LM | 971 020 | LHB | 971 030 | BFG | 971 043 | WW |
| 971 004 | LS | 971 012 | LM | 971 021 | DH | 971 031 | LS | 971 045 | LM |
| 971 006 | LM | 971 015 | LHB | 971 022 | BSE | 971 034 | LS | 971 051 | LS |
| 971 007 | LS | 971 016 | LL1 | 971 023 | LM | 971 035 | DH | 971 054 | LM |
| 971 008 | LS | 971 017 | BFG | 971 024 | LM | 971 038 | LHB | 971 055 | UEI |
| 971 009 | LM | 971 018 | LS | 971 027 | LM | 971 039 | LM | 971 056 | LM |
| 971 010 | LHB | 971 019 | LM | 971 028 | LM | 971 042 | WNT | 971 057 | LM |

| 971 059 | WP | 971 062 | DH | 971 066 | LM | 971 068 | LS | 971 070 | LS |
|---------|-----|---------|-----|---------|-----|---------|-----|---------|-----|
| 971 061 | BSE | 971 064 | BFG | 971 067 | DH | | | | |

## CLASS 971.6            Railbus Driving Trailer

**Built:** 1958-64 by Bautzen. Rebuilt 1992 onwards by DR/DBAG Halle Works.
**Wheel Arrangement:** 1-1.          **Seats:** 40S.
**Weight:**          **Maximum Speed:** 90 km/h.
**Length:** 13.550 m.
**Note:** Previously DR Class 171.6.

| 971 605 | WSR | 971 626 | WSR | 971 646 | WSR | 971 650 | WSR | 971 665 | WSR |
|---------|-----|---------|-----|---------|-----|---------|-----|---------|-----|
| 971 613 | WSR | 971 629 | WSR | 971 648 | LHB | 971 652 | WSR | 971 669 | WSR |
| 971 614 | WSR | 971 640 | LHB | 971 649 | UN | 971 658 | WSR | | |
| 971 625 | WSR | | | | | | | | |

## CLASSES 772.0 & 972.0          2-Car Railbus

**Built:** 1965 by Bautzen.
**Engine:** One Roßlau of 132 kW per power car.
**Transmission:** Mechanical. Strömungsmaschinen.
**Wheel Arrangement:** 1-A + 1-1.          **Seats:** 54S + 47S.
**Weight:** 19.40 + 14.00 tonnes.          **Maximum Speed:** 90 km/h.
**Length:** 13.550 + 13.550 m.
**Note:** Previously DR Classes 172.0 & 172.6.

| 772 001 | 972 601 | LS | 772 006 | 972 606 | WW | 772 011 | 972 611 | BSE |
|---------|---------|-----|---------|---------|-----|---------|---------|-----|
| 772 002 | 972 602 | LS | 772 007 | 972 607 | UN | 772 012 | 972 612 | LS |
| 772 003 | 972 603 | BCS | 772 008 | 972 608 | BCS | 772 014 | 972 614 | LS |
| 772 004 | 972 604 | WNT | 772 009 | 972 609 | BC | 772 015 | 972 615 | BCS |
| 772 005 | 972 605 | DH | 772 010 | 972 610 | LS | 772 016 | 972 616 | LS |

## CLASSES 772.1 & 972.1          2-Car Railbus

**Built:** 1968-69 by Görlitz.
**Engine:** One Roßlau of 132 kW per power car.
**Transmission:** Mechanical. Strömungsmaschinen.
**Wheel Arrangement:** 1-A + 1-1.          **Seats:** 54S + 47S (* 46S + 36S).
**Weight:** 22.10 + 14.00 tonnes.          **Maximum Speed:** 90 km/h.
**Length:** 13.550 + 13.550 m.
**Note:** Previously DR Classes 172.1 & 172.7.

| 772 101 | 972 701 | LL1 | 772 125 | 972 725 | BCS | 772 150 | 972 750 | WW |
|---------|---------|------|---------|---------|------|---------|---------|------|
| 772 102 | 972 702 | UN | 772 126 | 972 726 | WW | 772 151 | 972 751 | LS |
| 772 103 | 972 703 | DC | 772 127 | 972 727 | LL1 | 772 152 | 972 752 | DC |
| 772 104 | 972 704 | UN | 772 128 | 972 728 | LS | 772 153 | 972 753 | BPKR |
| 772 105 | 972 705 | BPKR | 772 129 | 972 729 | WW | 772 154 | 972 754 | LL1 |
| 772 106 | 972 706 | UN | 772 130 | 972 730 | LL1 | 772 155 | 972 755 | WW |
| 772 107 | 972 707 | BCS | 772 131* | 972 731 | BWUR | 772 156 | 972 756 | LHB |
| 772 108 | 972 708 | LHB | 772 132 | 972 732 | WW | 772 157 | 972 757 | UN |
| 772 109 | 972 709 | LL1 | | 972 734 | LL1 | 772 158 | 972 758 | WNT |
| 772 110 | 972 710 | WW | 772 135 | 972 735 | UN | 772 159 | 972 759 | UN |
| 772 111 | 972 711 | UN | 772 136 | 972 736 | LHB | 772 160 | 972 760 | LS |
| 772 112 | 972 712 | WNT | 772 137 | 972 737 | WW | 772 161 | 972 761 | WW |
| 772 113 | 972 713 | UN | 772 138 | 972 738 | LHB | 772 162 | 972 762 | UN |
| 772 114 | 972 714 | WW | 772 139 | 972 739 | UN | 772 163 | 972 763 | LL1 |
| 772 115 | 972 715 | LHB | 772 140 | 972 740 | UN | 772 164 | 972 764 | UN |
| 772 116 | 972 716 | WNT | 772 141 | 972 741 | UN | 772 165 | 972 765 | UN |
| 772 117 | 972 717 | LHB | 772 142 | 972 742 | BKR | 772 166 | 972 766 | LHB |
| 772 118 | 972 718 | WW | 772 143 | 972 743 | WW | 772 167 | 972 767 | UN |
| 772 119 | 972 719 | LL1 | 772 144 | 972 744 | LHB | 772 168 | 972 768 | LS |
| 772 120 | 972 720 | UN | 772 145 | 972 745 | DH | 772 169 | 972 769 | WW |
| 772 121 | 972 721 | DC | 772 146 | 972 746 | LL1 | 772 171 | 972 771 | UN |
| 772 122 | 972 722 | DC | 772 147 | 972 747 | WW | 772 172 | 972 772 | WNT |
| 772 123 | 972 723 | WW | 772 148 | 972 748 | LL1 | 772 173 | 972 773 | WW |
| 772 124 | 972 724 | WW | 772 149 | 972 749 | UN | | | |

# GERMANY

## CLASS 772.1                           Railbus

**Built:** 1958-64 by Bautzen as Class 971. Converted 1994 onwards by DBAG, Halle Works.
**Engine:** One Roßlau of 132 kW per power car.
**Transmission:** Mechanical. Strömungsmachinen.
**Wheel Arrangement:** 1-A               **Seats:**
**Weight:**                              **Maximum Speed:** 90 km/h.
**Length:** 13.550 m.

| | | | | |
|---|---|---|---|---|
| 772 174 WSR | 772 176 UN | 772 177 WW | 772 178 BSE | 772 179 UN |
| 772 175 WW | | | | |

## CLASS 772.3                           Railbus

**Built:** 1968-69 by Görlitz as Class 771.0. Rebuilt 1995 by DBAG, Halle Works.
**Engine:** One MAN.
**Transmission:** Hydraulic. Voith.
**Wheel Arrangement:** 1-A + 1-1.          **Seats**
**Weight:**                              **Maximum Speed:** 90 km/h.
**Length:** 13.550 m.

| | |
|---|---|
| 772 319 LM | 772 353 WW |

## CLASSES 772.4 &972.5              2-Car Railbus

**Built:** 1968-69 by Görlitz. Refurbished 1994 by DBAG, Halle Works.
**Engine:** One Roßlau of 132 kW per power car.
**Transmission:** Mechanical. Strömungsmachinen.
**Wheel Arrangement:** 1-A + 1-1.          **Seats:**
**Weight:**                              **Maximum Speed:** 90 km/h.
**Length:** 13.550 + 13.550 m.

772 413    972 513    BCS

## CLASS 786                           Single Unit

**Built:** 1935-36 by Bautzen.
**Engine:** One of 99 kW.
**Transmission:** Mechanical.
**Wheel Arrangement:** A-1.              **Seats:** 45S.
**Weight:** 17.70 tonnes.               **Maximum Speed:** 70 km/h.
**Length:** 12.200 m.

786 257 LHB

## CLASS 796                               Railbus

**Built:** 1956-62 by Uerdingen/MAN/WMD as class 798. Converted 1989 onwards for one person operation.
**Engines:** Two Büssing of 110 kW each.
**Transmission:** Mechanical.
**Wheel Arrangement:** B.               **Seats:** 56S.
**Weight:** 20.90 tonnes.               **Maximum Speed:** 90 km/h.
**Length:** 13.950 m.
**Note:** Operate with Class 996 trailers.

| | | | | |
|---|---|---|---|---|
| 796 597 FK | 796 724 NN1 | 796 757 TT | 796 784 ESIE | 796 816 TT |
| 796 625 TT | 796 739 TT | 796 760 ESIE | 796 785 ESIE | 796 825 NN1 |
| 796 690 ESIE | 796 740 TT | 796 761 FK | 796 802 FG | 796 828 TT |
| 796 702 ESIE | | | | |

## CLASS 996.0                       Railbus Trailer

**Built:** 1955-62 by Uerdingen/MAN/WMD as class 998. Converted 1989 onwards for one person operation.
**Wheel Arrangement:** 1-1.              **Seats:** 40S or 63S.

**Weight:** 10.30 or 10.70 tonnes.  
**Length:** 13.950 m.  
**Maximum Speed:** 90 km/h.

| | | | | |
|---|---|---|---|---|
| 996 225 TT | 996 248 TT | 996 286 TT | 996 298 TT | 996 299 TT |

## CLASS 996.6 — Railbus Driving Trailer

**Built:** 1955-62 by Uerdingen/MAN/WMD as class 998. Converted 1989 onwards for one person operation.  
**Wheel Arrangement:** 1-1.　**Seats:** 40S.  
**Weight:** 10.70 tonnes.　**Maximum Speed:** 90 km/h.  
**Length:** 13.950 m.

| | | | | |
|---|---|---|---|---|
| 996 628 (Z) | 996 683 FG | 996 739 TT | 996 766 ESIE | 996 782 TT |
| 996 641 TT | 996 701 ESIE | 996 742 TT | 996 768 ESIE | 996 783 TT |
| 996 658 ESIE | 996 702 (Z) | 996 747 TT | 996 773 FG | 996 791 (Z) |
| 996 661 FGO | 996 725 TT | 996 748 ESIE | 996 778 TT | 996 804 TT |
| 996 664 TT | 996 726 TT | 996 751 FG | 996 780 ESIE | 996 919 (Z) |
| 996 677 TT | | | | |

## CLASS 798 — Railbus

**Built:** 1955-62 by Uerdingen/MAN/WMD.  
**Engines:** Two Büssing of 110 kW each.  
**Transmission:** Mechanical.  
**Wheel Arrangement:** B.　**Seats:** 56S (* 52S).  
**Weight:** 20.90 tonnes.　**Maximum Speed:** 90 km/h.  
**Length:** 13.950 m.

| | | | | |
|---|---|---|---|---|
| 798 538 MMF | 798 610 MMF | 798 644 TT | 798 701 RK | 798 766 MMF |
| 798 565 TT | 798 622 RK | 798 647 MMF | 798 704 (Z) | 798 776 RO |
| 798 582 (Z) | 798 623 RK | 798 652* MMF | 798 726 TT | 798 818 RO |
| 798 589 FG | 798 629 FG | 798 653* MMF | 798 729 MMF | 798 823 ESIE |
| 798 598 FG | 798 632 MMF | 798 658 FG | 798 737 (Z) | |

## CLASS 997 — Railbus Driving Trailer

**Built:** 1959 by Uerdingen.  
**Wheel Arrangement:** 1-1.　**Seats:** 40S.  
**Weight:** 10.70 tonnes.　**Maximum Speed:** 90 km/h.  
**Length:** 13.950 m.

| | |
|---|---|
| 997 604 RO | 997 605 RO |

## CLASS 998.0 — Railbus Trailer

**Built:** 1955-62 by Uerdingen/MAN/WMD.  
**Wheel Arrangement:** 1-1.　**Seats:** 40S or 63S.  
**Weight:** 10.30 or 10.70 tonnes.　**Maximum Speed:** 90 km/h.  
**Length:** 13.950 m.

| | | | | |
|---|---|---|---|---|
| 998 031 TT | 998 064 MMF | 998 112 TT | 998 140 FG | 998 271 MMF |
| 998 045 RO | 998 069 TT | 998 133 (Z) | 998 142 FG | 998 318 MMF |
| 998 053 SKL | | | | |

## CLASS 998.6 — Railbus Driving Trailer

**Built:** 1955-62 by Uerdingen/MAN/WMD.  
**Wheel Arrangement:** 1-1.　**Seats:** 40S.  
**Weight:** 10.70 tonnes.　**Maximum Speed:** 90 km/h.  
**Length:** 13.950 m.

| | | | | |
|---|---|---|---|---|
| 998 642 MMF | 998 831 (Z) | 998 840 (Z) | 998 864 ESIE | 998 877 MMF |
| 998 654 (Z) | 998 834 FG | 998 850 FG | 998 867 ESIE | 998 878 FG |
| 998 744 (Z) | 998 836 MMF | 998 859 FG | 998 872 MMF | 998 880 FG |
| 998 786 TT | 998 838 FG | 998 863 ESIE | 998 874 MMF | 998 896 MMF |
| 998 801 TT | | | | |

# PORTUGAL

## VISAS

Not required by UK citizens (for stays of up to 2 months). Full UK Passport neccessary. British Visitors Passport not valid after 31.12.95.

## BRITISH EMBASSY

**Address:** Rue de S Domingos à Lapa 35-37, 37 1200 Lisbon.(Tel: (1) 396 1191, (1) 396 1147 or (1) 396 3181; FAX: (1) 397-6768).
There are British Consulates in Oporto and Portimão.

## CURRENCY & BANKING

Escudo (Esc). 1 Escudo = 100 centavos. Notes are in denominations of Esc 10,000, 5,000, 1,000, 500, 100 and 50. The 1,000 Escudo note is known as a "Conto". Coins are in denominations of Esc 100, 50, 20, 10, 5, 2.50 and 1, and 50 centavos.

Banking hours are usually 0830-1500 Monday to Friday only. Certain banks in Lisbon are also open 1800-2300 Monday to Friday. In the Algarve, the bank in the Vilamoura Marina Shopping Centre is open daily 0900-2100.

Visa, Access/Mastercard and American Express cards are widely accepted. Travellers Cheques are readily exchanged, and Eurocheques may be cashed at many banks.

## OTHER USEFUL INFORMATION

**Language:** Portuguese.

**Time:** GMT + 1 hour (GMT + 2 hours in summer).

**Electricity:** 220 V ac, 50 Hz. 110 Volts in some areas and 220 V dc in parts of the south. Continental 2-pin plugs are standard.

**Usual Public Holidays:** New Year's Day, Carnival Day (Feb 23), Good Friday, Liberty Day (Apr 25), May Day, Corpus Christi, Assumption, Republic Day (Oct 5), All Saints' Day, Portuguese Independence Day (Dec 1), Immaculate Conception Day (Dec 8).

Full details of these and other local holidays may be obtained from the Portuguese National Tourist Office.

## TOURIST INFORMATION

Portuguese National Tourist Office, 4th Floor, 22/25a Sackville Street, London, W1X 1DE. (Tel: 0171-494-1441; FAX 0171-494-1868).

## ACKNOWLEDGEMENT

Thanks are due to Philip Wormald for his help in compiling this section.

# Caminhos de Ferro Portugueses (CP).

**Gauges:** 1668 mm; 1000 mm.
**Route Length:** 2761 km (1668 mm), 307 km (1000 mm).
**Electrification System:** 25 kV 50 Hz ac overhead.

## NUMBERING

Although all vehicles (except Estoril Line) carry the 12 digit UIC style numbers introduced in 1974, numbers issued in the former series are still retained and it is these numbers which are used in this publication, although details of digits 5 to 11 of the longer numbers are given in class headings to assist readers with identification.

# DEPOTS

The following depots have an allocation and are shown together with the codes used to denote them in this book. All depots shown are 1668 mm gauge and on the main system unless noted otherwise.

| | | | |
|---|---|---|---|
| BA | Barreiro | MI | Mirandela (1000 mm gauge) |
| CO | Coimbra B | PA | Porto Campanhã |
| EN | Entroncamento | PB | Porto Boa Vista (1000 mm gauge) |
| FI | Figueira da Foz | PC | Porto Contumil |
| LC | Lisboa Cais do Sodré | RE | Régua (1000 mm gauge) |
| LI | Livraçao (1000 mm gauge) | SE | Sernada do Vouga (1000 mm gauge) |
| LR | Lisboa Campolide | | |

# OTHER DEPOTS & STABLING POINTS

Lisboa Santa Apolónia, Pamphilosa.

# WORKSHOPS

**Barreiro** - 1668 mm gauge diesel locomotives.
**Custoias (Porto)** - 1000 mm gauge locomotives and rolling stock. Certain 1668 mm gauge DMUs.
**Entroncamento** - Electric locomotives, DMUs, EMUs, coaches and wagons.
**Figueira da Foz** - 1151 Class locomotives, coaches.
Estoril line stock is fully maintained at Lisboa Cais do Sodré.

# RAILROVER INFORMATION

**Bilhete Turistico**
7, 14 or 21 consecutive days unlimited 2nd Class travel on all CP services (trains, buses and ferries). Supplements/Reservations are applicable for travel on Rápido Alfa, Alfa and Rápido Intercidades services. Passport required.
Prices:    7 Days   ESC 17,000
               14 Days  ESC 27,000
               21 Days  ESC 38,000

**Portugal Explorer**
Available to holders of an International Student Card or persons aged under 26 only. 7, 14 or 21 consecutive days unlimited 2nd Class travel on all CP services (trains, buses and ferries). Travel must be completed not later than 31 October of year of issue.
Prices:    7 Days   £ 48.00
               14 Days  £ 81.00
               21 Days  £120.00

For details of Freedom Pass (Euro Domino) tickets, please see the section at the front of this book.

**Tickets & Details**
Bilhete Turistico are available from major CP stations.
Portugal Explorer tickets are available from Eurotrain/UIST, 52 Grosvenor Gardens, Victoria, London, SW1W 0AG (Tel: 0171-730-3402).

# TIMETABLE

The "Guia Horário Oficial" is published twice a year and is available from main CP stations. It is also available in the UK from European Rail Timetables (Tel: 01909-485855), 39 Kilton Glade, Worksop, Nottinghamshire, S81 0PX (1995 Summer price £6.25, including postage & packing).

# SPECIAL NOTE

Updated information on CP multiple units is extremely difficult to come by, and readers are invited to send in details of their observations to update future editions. The information in this book has been compiled on the basis of such observations.

# PORTUGAL

# 1668 mm GAUGE DIESEL LOCOMOTIVES

## 1001-06 C

**Built:** 1948 by Drewry.
**Engine:** Gardner 8L3-8 of 147 kW at 1200 rpm.
**Transmission:** Mechanical.
**Weight:** 30.40 tonnes.                    **Length:** 7.815 m.
**Maximum Speed:** 41.5 km/h.              **Train Supply:** Not equipped.
**UIC Numbers:** 2 011001 etc.

| | | | | | | | | | |
|---|---|---|---|---|---|---|---|---|---|
| 1001 | PC | 1002 | PC | 1003 | PC | 1005 | PC | 1006 | PC |

## 1051-65 B

**Built:** 1955 by Moyse.
**Engine:** Moyse M6A-6 of 270 kW at 1500 rpm.
**Transmission:** Electric. Moyse.
**Weight:** 28.30 tonnes.                    **Length:** 7.280 m.
**Maximum Speed:** 38 km/h.                **Train Supply:** Not equipped.
**UIC Numbers:** 2 011051 etc.

| | | | | | | | | | |
|---|---|---|---|---|---|---|---|---|---|
| 1051 | EN | 1054 | EN | 1057 | EN | 1060 | BA | 1064 | EN |
| 1052 | EN | 1055 | BA | 1059 | EN | 1062 | EN | 1065 | BA |
| 1053 | BA | 1056 | EN | | | | | | |

## 1101-11 Bo-Bo

**Built:** 1949 by General Electric.
**Engines:** Two Caterpillar D 17000-8 of 141 kW each at 1000 rpm.
**Transmission:** Electric. General Electric.
**Weight:** 41.20 tonnes.                    **Length:** 10.210 m.
**Maximum Speed:** 56 km/h.                **Train Supply:** Not equipped.
**Note:** 1109 was previously 1103.
**UIC Numbers:** 2 021101 etc.

| | | | | | | | | | |
|---|---|---|---|---|---|---|---|---|---|
| 1101 | BA | 1104 | BA | 1106 | BA | 1108 | BA | 1110 | BA |
| 1102 | FI | 1105 | BA | 1107 | FI | 1109 | FI | 1111 | BA |

## 1151-86 C

**Built:** 1966-67 by SOREFAME.
**Engine:** Rolls Royce C8 TFL Mk4 of 258 kW at 1800 rpm.
**Transmission:** Hydraulic. Rolls Royce CF 11 500.
**Weight:** 42.00 tonnes.                    **Length:** 8.517 m.
**Maximum Speed:** 58 km/h.                **Train Supply:** Not equipped.
**UIC Numbers:** 2 021151 etc.

| | | | | | | | | | |
|---|---|---|---|---|---|---|---|---|---|
| 1151 | EN | 1159 | LR | 1166 | LR | 1173 | LR | 1180 | PC |
| 1152 | EN | 1160 | LR | 1167 | LR | 1174 | CO | 1181 | PC |
| 1153 | EN | 1161 | LR | 1168 | LR | 1175 | CO | 1182 | PC |
| 1154 | EN | 1162 | LR | 1169 | LR | 1176 | CO | 1183 | PC |
| 1155 | EN | 1163 | LR | 1170 | LR | 1177 | CO | 1184 | PC |
| 1156 | EN | 1164 | LR | 1171 | LR | 1178 | CO | 1185 | PC |
| 1157 | EN | 1165 | LR | 1172 | LR | 1179 | CO | 1186 | PC |
| 1158 | EN | | | | | | | | |

## 1201-25 Bo-Bo

**Built:** 1961-64 by SOREFAME.
**Engine:** SACM MGO V12 ASHR of 611 kW at 1500 rpm.
**Transmission:** Electric. B & L.
**Weight:** 64.70 tonnes.                    **Length:** 14.680 m.

**Maximum Speed:** 80 Km/h.
**UIC Numbers:** 1 061201 etc.

**Train Supply:** Not equipped.

| | | | | | | | | | |
|---|---|---|---|---|---|---|---|---|---|
| 1201 | LR | 1206 | BA | 1211 | BA | 1216 | BA | 1221 | BA |
| 1202 | LR | 1207 | BA | 1212 | BA | 1217 | BA | 1222 | BA |
| 1203 | LR | 1208 | BA | 1213 | BA | 1218 | BA | 1223 | BA |
| 1204 | LR | 1209 | BA | 1214 | BA | 1219 | BA | 1224 | BA |
| 1205 | BA | 1210 | BA | 1215 | BA | 1220 | BA | 1225 | BA |

## 1321-38 Co-Co

**Built:** 1965-67 by Euskalduna. Acquired 1989-1990 from Red Nacional de los Ferrocarriles Españoles and rebuilt by ATIENSA.
**Engine:** Alco 251D-6 of 1007 kw at 1100 rpm.
**Transmission:** Electric. General Electric.
**Weight:** 83.91 tonnes.
**Maximum Speed:** 120 km/h.
**UIC Numbers:** 1 101321 etc.

**Length:** 16.237 m.
**Train Supply:** Not equipped.

| | | | | |
|---|---|---|---|---|
| 1321 | (RENFE 313.011) BA | | 1331 | (RENFE 313.030) BA |
| 1322 | (RENFE 313.039) BA | | 1332 | (RENFE 313.029) BA |
| 1323 | (RENFE 313.001) BA | | 1333 | (RENFE 313.028) BA |
| 1325 | (RENFE 313.007) BA | | 1334 | (RENFE 313.017) BA |
| 1326 | (RENFE 313.002) BA | | 1335 | (RENFE 313.034) BA |
| 1327 | (RENFE 313.035) BA | | 1336 | (RENFE 313.032) BA |
| 1328 | (RENFE 313.014) BA | | 1337 | (RENFE 313.015) BA |
| 1329 | (RENFE 313.020) BA | | 1338 | (RENFE 313.037) BA |
| 1330 | (RENFE 313.008) BA | | | |

## 1401-1467 Bo-Bo

**Built:** 1967-69 by English Electric/SOREFAME.
**Engine:** English Electric 8CSVT of 984 kW at 850 rpm.
**Transmission:** Electric. English Electric.
**Weight:** 64.40 tonnes.
**Maximum Speed:** 105 km/h.
**UIC Numbers:** 1 101401 etc.

**Length:** 12.720 m.
**Train Supply:** Not equipped.

| | | | | | | | | | |
|---|---|---|---|---|---|---|---|---|---|
| 1401 | PC | 1415 | PC | 1428 | PC | 1444 | CO | 1456 | EN |
| 1402 | PC | 1416 | PC | 1429 | PC | 1445 | CO | 1457 | EN |
| 1404 | PC | 1417 | PC | 1431 | PC | 1446 | CO | 1458 | EN |
| 1405 | PC | 1418 | PC | 1432 | PC | 1447 | CO | 1459 | EN |
| 1406 | PC | 1419 | PC | 1434 | PC | 1448 | CO | 1460 | EN |
| 1407 | PC | 1421 | PC | 1435 | PC | 1449 | CO | 1461 | EN |
| 1408 | PC | 1422 | PC | 1436 | CO | 1450 | CO | 1462 | EN |
| 1409 | PC | 1423 | PC | 1437 | CO | 1451 | CO | 1463 | EN |
| 1410 | PC | 1424 | PC | 1438 | PC | 1452 | CO | 1464 | EN |
| 1411 | PC | 1425 | PC | 1440 | CO | 1453 | CO | 1465 | PC |
| 1412 | PC | 1426 | PC | 1441 | PC | 1454 | CO | 1466 | PC |
| 1413 | PC | 1427 | PC | 1442 | CO | 1455 | CO | 1467 | PC |
| 1414 | PC | | | | | | | | |

## 1501-12/21-25* A1A-A1A

**Built:** 1948 (*1955) by Alco. Re-engined at CP Barreiro Works 1971-78.
**Engine:** Alco 251C-12 of 1618 kW at 1025 rpm.
**Transmission:** Electric. General Electric.
**Weight:** 111.00 (* 114.00) tonnes.
**Maximum Speed:** 120 km/h.
**UIC Numbers:** 1 171501 etc.

**Length:** 16.988 m.
**Train Supply:** Not equipped.

| | | | | | | | | | |
|---|---|---|---|---|---|---|---|---|---|
| 1501 | BA | 1505 | BA | 1509 | BA | 1512 | BA | 1523* | BA |
| 1502 | BA | 1506 | BA | 1510 | BA | 1521* | BA | 1524* | BA |
| 1503 | BA | 1507 | BA | 1511 | BA | 1522* | BA | 1525* | BA |
| 1504 | BA | 1508 | BA | | | | | | |

# PORTUGAL

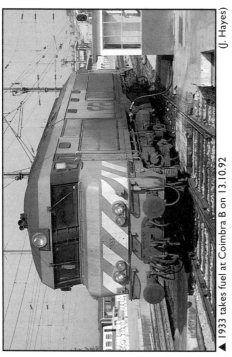

◄ 1933 takes fuel at Coimbra B on 13.10.92 (J. Hayes)

▼ Electrics of Classes 2501 & 2601 at Lisboa Santa Apolónia in May 1993 (B. Philpott)

◄ 1421 at Tua during September 1993 (B. Philpott)

▼ 1552 at Abrantes on 06.10.93 (J. Hayes)

# 1551-70      Co-Co

**Built:** 1973 by MLW.
**Engine:** MLW/Alco 251C3-12 of 1615 kW at 1050 rpm.
**Transmission:** Electric. General Electric.
**Weight:** 89.70 tonnes.      **Length:** 17.905 m.
**Maximum Speed:** 120 km/h.      **Train Supply:** Not equipped.
**UIC Numbers:** 1 171551 etc.

| | | | | | | | | | |
|---|---|---|---|---|---|---|---|---|---|
| 1551 | EN | 1555 | EN | 1559 | EN | 1563 | EN | 1567 | EN |
| 1552 | EN | 1556 | EN | 1560 | EN | 1564 | EN | 1568 | EN |
| 1553 | EN | 1557 | EN | 1561 | EN | 1565 | EN | 1569 | EN |
| 1554 | EN | 1558 | EN | 1562 | EN | 1566 | EN | 1570 | EN |

# 1801-10      Co-Co

**Built:** 1968-69 by English Electric.
**Engine:** English Electric 16CSVT of 1940 kW at 850 rpm.
**Transmission:** Electric. English Electric.
**Weight:** 110.30 tonnes.      **Length:** 18.680 m.
**Maximum Speed:** 140 km/h.      **Train Supply:** Not equipped.
**UIC Numbers:** 1 201801 etc.

| | | | | | | | | | |
|---|---|---|---|---|---|---|---|---|---|
| 1801 | BA | 1803 | BA | 1805 | BA | 1807 | BA | 1810 | BA |
| 1802 | BA | 1804 | BA | 1806 | BA | 1808 | BA | | |

# 1901-13      Co-Co

**Built:** 1981 by SOREFAME.
**Engine:** SACM AGO V12 DSHR of 2220 kW.
**Transmission:** Electric. Alsthom.
**Weight:** 117.00 tonnes.      **Length:** 19.084 m.
**Maximum Speed:** 100 km/h.      **Train Supply:** Not equipped.
**UIC Numbers:** 1 221901 etc.

| | | | | | | | | | |
|---|---|---|---|---|---|---|---|---|---|
| 1901 | BA | 1904 | BA | 1907 | BA | 1910 | BA | 1912 | BA |
| 1902 | BA | 1905 | BA | 1908 | BA | 1911 | BA | 1913 | BA |
| 1903 | BA | 1906 | BA | 1909 | BA | | | | |

# 1931-47      Co-Co

**Built:** 1981 by SOREFAME.
**Engine:** SACM AGO V12 DSHR of 2220 kW.
**Transmission:** Electric. Alsthom.
**Weight:** 116.50 tonnes.      **Length:** 18.756 m.
**Maximum Speed:** 120 km/h.      **Train Supply:** Electric.
**UIC Numbers:** 1 221931 etc.

| | | | | | | | | | |
|---|---|---|---|---|---|---|---|---|---|
| 1931 | EN | 1935 | EN | 1939 | EN | 1942 | EN | 1945 | EN |
| 1932 | EN | 1936 | EN | 1940 | EN | 1943 | EN | 1946 | EN |
| 1933 | EN | 1937 | EN | 1941 | EN | 1944 | EN | 1947 | EN |
| 1934 | EN | 1938 | EN | | | | | | |

**Name:** 1935 *Cidade do Barreiro*

# 1962-73      Co-Co

**Built:** 1973 by Bombardier.
**Engine:** Bombardier/Alco 251E-16 of 2250 kW.
**Transmission:** Electric. General Electric.
**Weight:** 121.00 tonnes.      **Length:** 19.895 m.
**Maximum Speed:** 120 km/h.      **Train Supply:** Electric.
**UIC Numbers:** 1 221962 etc.

| | | | | | | | | | |
|---|---|---|---|---|---|---|---|---|---|
| 1962 | FI | 1965 | FI | 1968 | FI | 1970 | FI | 1972 | FI |
| 1963 | FI | 1966 | FI | 1969 | FI | 1971 | FI | 1973 | FI |
| 1964 | FI | 1967 | FI | | | | | | |

# PORTUGAL

## 1668 mm GAUGE ELECTRIC LOCOMOTIVES

### 2501-15 Bo-Bo

**Built:** 1956-57 by 50 Hz Groupement.
**Electrical Equipment:** Siemens/Alsthom/Schneider-Jeumont.
**Continuous Rating:** 2116 kW.
**Weight:** 72.00 tonnes.　　**Length:** 15.380 m.
**Maximum Speed:** 120 km/h.　　**Train Supply:** Electric.
**UIC Numbers:** 0 272501 etc.

| | | | | | | | | | |
|---|---|---|---|---|---|---|---|---|---|
| 2501 | EN | 2504 | EN | 2507 | EN | 2510 | EN | 2513 | EN |
| 2502 | EN | 2505 | EN | 2508 | EN | 2511 | EN | 2514 | EN |
| 2503 | EN | 2506 | EN | 2509 | EN | 2512 | EN | 2515 | EN |

### 2551-70 Bo-Bo

**Built:** 1963-64 by SOREFAME.
**Electrical Equipment:** Siemens/Alsthom/Schneider-Jeumont.
**Continuous Rating:** 2116 kW.
**Weight:** 70.50 tonnes.　　**Length:** 15.380 m.
**Maximum Speed:** 120 km/h.　　**Train Supply:** Electric.
**UIC Numbers:** 0 272551 etc.

| | | | | | | | | | |
|---|---|---|---|---|---|---|---|---|---|
| 2551 | EN | 2557 | EN | 2561 | EN | 2565 | EN | 2568 | EN |
| 2552 | EN | 2558 | EN | 2562 | EN | 2566 | EN | 2569 | EN |
| 2553 | EN | 2559 | EN | 2563 | EN | 2567 | EN | 2570 | EN |
| 2556 | EN | 2560 | EN | 2564 | EN | | | | |

### 2601-12 Bo-Bo

**Built:** 1974 by Alsthom.
**Electrical Equipment:** Siemens/Alsthom/Schneider-Jeumont.
**Continuous Rating:** 2940 kW.
**Weight:** 78.00 tonnes.　　**Length:** 17.500 m.
**Maximum Speed:** 160 km/h.　　**Train Supply:** Electric.
**UIC Numbers:** 0 382601 etc.

| | | | | | | | | | |
|---|---|---|---|---|---|---|---|---|---|
| 2601 | EN | 2604 | EN | 2607 | EN | 2609 | EN | 2611 | EN |
| 2602 | EN | 2605 | EN | 2608 | EN | 2610 | EN | 2612 | EN |
| 2603 | EN | 2606 | EN | | | | | | |

### 2621-29 Bo-Bo

**Built:** 1987 by SOREFAME.
**Electrical Equipment:** Siemens/Alsthom/Schneider-Jeumont.
**Continuous Rating:** 2940 kW.
**Weight:** 78.00 tonnes.　　**Length:** 17.500 m.
**Maximum Speed:** 160 km/h.　　**Train Supply:** Electric.
**UIC Numbers:** 0 392621 etc.

| | | | | | | | | | |
|---|---|---|---|---|---|---|---|---|---|
| 2621 | EN | 2623 | EN | 2625 | EN | 2627 | EN | 2629 | EN |
| 2622 | EN | 2624 | EN | 2626 | EN | 2628 | EN | | |

### 5601-30 Bo-Bo

**Built:** 1993-95 by SOREFAME.
**Electrical Equipment:** Siemens.
**Continuous Rating:** 5600 kW.
**Weight:** 88.00 tonnes.　　**Length:** 20.380 m.
**Maximum Speed:** 220 km/h.　　**Train Supply:** Electric.
**UIC Numbers:** 0 755601 etc.

| | | | | |
|---|---|---|---|---|
| 5601 EN | 5607 EN | 5613 EN | 5619 EN | 5625 EN |
| 5602 EN | 5608 EN | 5614 EN | 5620 EN | 5626 EN |
| 5603 EN | 5609 EN | 5615 EN | 5621 EN | 5627 EN |
| 5604 EN | 5610 EN | 5616 EN | 5622 EN | 5628 EN |
| 5605 EN | 5611 EN | 5617 EN | 5623 EN | 5629 EN |
| 5606 EN | 5612 EN | 5618 EN | 5624 EN | 5630 EN |

**Name:** 5601 *Entroncamento*

## 1668 mm GAUGE DIESEL MULTIPLE UNITS

## CLASS 0101      Power Car

**Built:** 1948 by Nohab.
**Normal Formation:** AB.
**Engines:** Two Saab-Scania of 93 kW each.
**Length:** 22.490 m.
**Weight:**
**Note:** Operate with Class 0101 trailers (see below).
**UIC Numbers:** 7 020101 etc.

**Seats:** 16F, 56S.
**Transmission:** Hydraulic. Voith.
**Wheel Arrangement:** 1A-A1.
**Maximum Speed:** 100 km/h.

| | | | | |
|---|---|---|---|---|
| 0101 BA | 0107 BA | 0110 BA | 0112 BA | 0114 BA |
| 0105 BA | 0108 BA | 0111 BA | 0113 BA | 0115 BA |
| 0106 BA | 0109 BA | | | |

## CLASS 0101      Trailer Car

**Built:** 1948 by Nohab.
**Normal Formation:** AB (* B)
**Length:** 22.490 m.
**Weight:**
**Note:** Operate with Class 0101 power cars (see above).
**UIC Numbers:** 37 29101 etc.

**Seats:** 16F, 38S (* 58S).
**Wheel Arrangement:** 2-2.
**Maximum Speed:** 100 km/h.

| | | | | |
|---|---|---|---|---|
| 9101* BA | 9103 BA | 9104* BA | 9108 BA | 9111 BA |
| 9102 BA | | | | |

## CLASS 0301      Power Car

**Built:** 1954 by Allan.
**Normal Formation:** AB.
**Engines:** Two SSCM of 133 kW each.
**Length:** 23.630 m.
**Weight:** 51.50 tonnes.
**Note:** Operate with Class 0301 trailers (see below).
**UIC Numbers:** 8 030301 etc.

**Seats:** 24F, 50S.
**Transmission:** Electric. Smit.
**Wheel Arrangement:** Bo-Bo.
**Maximum Speed:** 100 km/h.

| | | | | |
|---|---|---|---|---|
| 0301 CO | 0306 CO | 0313 CO | 0318 CO | 0322 CO |
| 0302 CO | 0307 CO | 0314 CO | 0319 CO | 0323 CO |
| 0303 CO | 0308 CO | 0315 CO | 0320 CO | 0324 CO |
| 0304 CO | 0310 CO | 0316 CO | 0321 CO | 0325 CO |
| 0305 CO | 0312 CO | 0317 CO | | |

## CLASS 0301      Trailer Car

**Built:** 1954 by Allan.
**Normal Formation:** B.
**Length:** 23.630 m.
**Weight:** 31.50 tonnes.
**Note:** Operate with Class 0301 power cars (see above).
**UIC Numbers:** 22 29301 etc.

**Seats:** 106S.
**Wheel Arrangement:** 2-2.
**Maximum Speed:** 100 km/h.

| | | | | |
|---|---|---|---|---|
| 9301 CO | 9304 (Z) | 9306 CO | 9308 CO | 9312 CO |
| 9303 (Z) | 9305 (Z) | 9307 CO | 9310 CO | |

PORTUGAL

## CLASS 0401 — 2-Car Unit

**Built:** 1965-66 by SOREFAME.
**Normal Formation:** B + AB.    **Seats:**
**Engines:** Two Rolls Royce of 207 kW each per power car.
**Transmission:** Hydraulic. Rolls Royce.
**Length:** 51.960 m.    **Wheel Arrangement:** 1A-A1 + 2-2.
**Weight:** 94.10 tonnes.    **Maximum Speed:** 110 km/h.
**UIC Numbers:** 7 050401 + 5 000401 etc.

| | | | | |
|---|---|---|---|---|
| 0401 PA | 0405 PA | 0409 PA | 0413 PA | 0417 PA |
| 0402 PA | 0406 PA | 0410 PA | 0414 PA | 0418 PA |
| 0403 PA | 0407 PA | 0411 PA | 0415 PA | 0419 PA |
| 0404 PA | 0408 PA | 0412 PA | 0416 PA | |

## CLASS 0601 — 3-Car Unit

**Built:** 1979 by SOREFAME. Originally 2-Car units. Strengthened 1989-90.
**Normal Formation:** AB + B + B.    **Seats:** 40F, 270S.
**Engines:** Two SFAC of 288 kW each per power car.
**Transmission:** Hydraulic. Voith.
**Length:** 80.220 m.    **Wheel Arrangement:** 2-B + 2-2 + B-2.
**Weight:**    **Maximum Speed:** 120 km/h.
**Note:** Trailers do not remain constant, and may be mixed with Class 0651.
**UIC Numbers:** 7 030601 + 5 990601 + 7 030621 etc.

| | | | | |
|---|---|---|---|---|
| 0601 PA | 0605 PA | 0609 PA | 0613 PA | 0617 PA |
| 0602 PA | 0606 PA | 0610 PA | 0614 PA | 0618 PA |
| 0603 PA | 0607 PA | 0611 PA | 0615 PA | 0619 PA |
| 0604 PA | 0608 PA | 0612 PA | 0616 PA | 0620 PA |

## CLASS 0651 — 3-Car Unit

**Built:** 1989-90 by SOREFAME.
**Normal Formation:** AB + B + B.    **Seats:** 40F, 270S.
**Engines:** Two SFAC of 288 kW each per power car.
**Transmission:** Hydraulic. Voith.
**Length:** 80.220 m.    **Wheel Arrangement:** 2-B + 2-2 + B-2.
**Weight:**    **Maximum Speed:** 120 km/h.
**Note:** Trailers do not remain constant, and may be mixed with Class 0601.
**UIC Numbers:** 7 030651 + 5 990651 + 7 030657 etc.

| | | | | |
|---|---|---|---|---|
| 0651 FI | 0653 FI | 0654 FI | 0655 FI | 0656 FI |
| 0652 FI | | | | |

# 1668 mm GAUGE ELECTRIC MULTIPLE UNITS

## CLASS 2001 — 3-Car Unit

**Built:** 1956 by SOREFAME.
**Normal Formation:** B + A + B.    **Seats:** 68F, 176S.
**Electrical Equipment:** Siemens/AEG/Oerlikon.    **Length:** 71.060 m.
**Motors:** 4 of 272 kW each.    **Wheel Arrangement:** Bo-Bo + 2-2 + 2-2.
**Weight:** 117.00 tonnes.    **Maximum Speed:** 90 km/h.
**UIC Numbers:** 9 142001 + 5 992001 + 5 002001 etc.

| | | | | |
|---|---|---|---|---|
| 2001 LR | 2006 LR | 2012 LR | 2017 LR | 2021 LR |
| 2002 LR | 2007 LR | 2013 LR | 2018 LR | 2022 LR |
| 2003 LR | 2008 LR | 2014 LR | 2019 LR | 2023 LR |
| 2004 LR | 2010 LR | 2015 LR | 2020 LR | 2025 LR |
| 2005 LR | | | | |

# CLASS 2051        3-Car Unit

**Built:** 1962-66 by SOREFAME.
**Normal Formation:** B + AB + B.
**Electrical Equipment:** Siemens/AEG/Oerlikon.
**Motors:** 4 of 272 kW each.
**Weight:** 123.60 tonnes.
**UIC Numbers:** 9 142051 + 5 992051 + 5 002051 etc.

**Seats:** 56F, 218S.
**Length:** 71.060 m.
**Wheel Arrangement:** Bo-Bo + 2-2 + 2-2.
**Maximum Speed:** 90 km/h.

| | | | | | | | | | |
|---|---|---|---|---|---|---|---|---|---|
| 2051 | LR | 2058 | LR | 2065 | LR | 2072 | LR | 2085 | FI |
| 2052 | LR | 2059 | LR | 2066 | LR | 2073 | LR | 2086 | LR |
| 2053 | FI | 2060 | FI | 2067 | LR | 2074 | LR | 2087 | FI |
| 2054 | FI | 2061 | LR | 2068 | LR | 2082 | FI | 2088 | FI |
| 2055 | FI | 2062 | LR | 2069 | LR | 2083 | LR | 2089 | FI |
| 2056 | LR | 2063 | LR | 2070 | LR | 2084 | FI | 2090 | FI |
| 2057 | LR | 2064 | LR | 2071 | LR | | | | |

# CLASS 2101        3-Car Unit

**Built:** 1970 by SOREFAME.
**Normal Formation:** B + A + B.
**Electrical Equipment:** Siemens/AEG/Oerlikon.
**Motors:** 4 of 318 kW each.
**Weight:** 132.80 tonnes.
**UIC Numbers:** 5 002101 + 9 172101 + 5 002102 etc.

**Seats:** 60F, 159S.
**Length:** 71.060 m.
**Wheel Arrangement:** 2-2 + Bo-Bo + 2-2.
**Maximum Speed:** 120 km/h.

| | | | | | | | | | |
|---|---|---|---|---|---|---|---|---|---|
| 2101 | EN | 2106 | EN | 2111 | EN | 2116 | EN | 2121 | EN |
| 2102 | EN | 2107 | EN | 2112 | EN | 2117 | EN | 2122 | EN |
| 2103 | EN | 2108 | EN | 2113 | EN | 2118 | EN | 2123 | EN |
| 2104 | EN | 2109 | EN | 2114 | EN | 2119 | EN | 2124 | EN |
| 2105 | EN | 2110 | EN | 2115 | EN | 2120 | EN | | |

# CLASS 2151        3-Car Unit

**Built:** 1977 by SOREFAME.
**Normal Formation:** B + A + B.
**Electrical Equipment:** Siemens/AEG/Oerlikon.
**Motors:** 4 of 318 kW each.
**Weight:** 132.80 tonnes.
**UIC Numbers:** 5 002151 + 9 172151 + 5 002152 etc.

**Seats:** 60F, 192S.
**Length:** 71.060 m.
**Wheel Arrangement:** 2-2 + Bo-Bo + 2-2.
**Maximum Speed:** 120 km/h.

| | | | | | | | | | |
|---|---|---|---|---|---|---|---|---|---|
| 2151 | EN | 2155 | EN | 2159 | EN | 2163 | EN | 2166 | EN |
| 2152 | EN | 2156 | EN | 2160 | EN | 2164 | EN | 2167 | EN |
| 2153 | EN | 2157 | EN | 2161 | EN | 2165 | EN | 2168 | EN |
| 2154 | EN | 2158 | EN | 2162 | EN | | | | |

# CLASS 2201        3-Car Unit

**Built:** 1984 by SOREFAME.
**Normal Formation:** B + A + B.
**Electrical Equipment:** Siemens/AEG/Oerlikon.
**Motors:** 4 of 318 kW each.
**Weight:** 132.80 tonnes.
**UIC Numbers:** 5 002201 + 9 172201 + 5 002202 etc.

**Seats:** 60F, 192S.
**Length:** 71.060 m.
**Wheel Arrangement:** 2-2 + Bo-Bo + 2-2.
**Maximum Speed:** 120 km/h.

| | | | | | | | | | |
|---|---|---|---|---|---|---|---|---|---|
| 2201 | EN | 2204 | EN | 2207 | EN | 2210 | EN | 2213 | EN |
| 2202 | EN | 2205 | EN | 2208 | EN | 2211 | EN | 2214 | EN |
| 2203 | EN | 2206 | EN | 2209 | EN | 2212 | EN | 2215 | EN |

# CLASS 2301        4-Car Unit

**Built:** 1993 onwards by SOREFAME.
**Normal Formation:**
**Electrical Equipment:** Siemens.

**Seats:**
**Length:**

# PORTUGAL

◄ A Class 9301 railcar arrives at Aveiro with the 0903 from Sernada do Vouga on 20.05.93 (B. Philpott)

▼ Class 9401 3-car unit at Sernada do Vouga during May 1993 (B. Philpott)

◄ A 2001 Class emu at Lisboa Santa Apolónia (E. Barnes)

▼ 1000 mm gauge railcar 9102 at Porto (E. Barnes)

**Motors:** 8 of 288 kW each.
**Weight:**
**UIC Numbers:** 9 152301 + 5 002301 + 9 152302 etc.

**Wheel Arrangement:** Bo-Bo + 2-2 + 2-2 + Bo-Bo.
**Maximum Speed:** 120 km/h.

| | | | | | | | | | |
|---|---|---|---|---|---|---|---|---|---|
| 2301 | LR | 2310 | LR | 2319 | LR | 2327 | LR | 2335 | LR |
| 2302 | LR | 2311 | LR | 2320 | LR | 2328 | LR | 2336 | LR |
| 2303 | LR | 2312 | LR | 2321 | LR | 2329 | LR | 2337 | LR |
| 2304 | LR | 2313 | LR | 2322 | LR | 2330 | LR | 2338 | LR |
| 2305 | LR | 2314 | LR | 2323 | LR | 2331 | LR | 2339 | LR |
| 2306 | LR | 2315 | LR | 2324 | LR | 2332 | LR | 2340 | LR |
| 2307 | LR | 2316 | LR | 2325 | LR | 2333 | LR | 2341 | LR |
| 2308 | LR | 2317 | LR | 2326 | LR | 2334 | LR | 2342 | LR |
| 2309 | LR | 2318 | LR | | | | | | |

# 1000 mm GAUGE DIESEL LOCOMOTIVES

## 9001-06     Bo-Bo

**Built:** 1959-65 by Alsthom (9001-5); 1967 by Euskalduna (9006). Acquired from Ferrocarrill del Tajuña in 1975.
**Engine:** SACM MGO V12 ASHR of 630 (* 575) kW at 1500 rpm.
**Transmission:** Electric. Alsthom.
**Weight:** 46.00 (* 43.00) tonnes.
**Maximum Speed:** 70 km/h.
**UIC Numbers:** 1 069001 etc.

**Length:** 11.174 m.
**Train Supply:** Not equipped.

| | | | | | | | | | |
|---|---|---|---|---|---|---|---|---|---|
| 9001 | PB | 9003 | PB | 9004 | PB | 9005 | PB | 9006 | PB |
| 9002 | PB | | | | | | | | |

## 9021-31     Bo-Bo

**Built:** 1976-78 by Alsthom.
**Engine:** SACM MGO V12 ASHR of 740 kW at 1500 rpm.
**Transmission:** Electric. Alsthom.
**Weight:** 46.80 tonnes.
**Maximum Speed:** 70 km/h.
**UIC Numbers:** 1 079021 etc.

**Length:** 11.360 m.
**Train Supply:** Not equipped.

| | | | | | | | | | |
|---|---|---|---|---|---|---|---|---|---|
| 9021 | PB | 9024 | PB | 9026 | PB | 9028 | PB | 9030 | PB |
| 9022 | PB | 9025 | PB | 9027 | PB | 9029 | PB | 9031 | PB |
| 9023 | PB | | | | | | | | |

# 1000 mm GAUGE DIESEL MULTIPLE UNITS

## CLASS 9101     Single Unit

**Built:** 1949 by Nohab.
**Normal Formation:** AB.
**Engines:** Two Scania Vabis of 88 kW each.
**Length:** 15.500 m.
**Weight:** 22.00 tonnes.
**UIC Numbers:** 7 029101 etc.

**Seats:** 8F, 28S.
**Transmission:** Hydraulic. Lysholm-Smith.
**Wheel Arrangement:** B-B.
**Maximum Speed:** 70 km/h.

| | | | | | |
|---|---|---|---|---|---|
| 9101 | LI | 9102 | LI | 9103 | LI |

## CLASS 9301     Single Unit

**Built:** 1954 by Allan.
**Normal Formation:** AB.
**Engines:** Two Volvo.
**Length:** 19.510 m.

**Seats:** 12F, 32S.
**Transmission:** Electric. Smit.
**Wheel Arrangement:** Bo-Bo.

# PORTUGAL

**Weight:** 37.00 tonnes.
**Note:** Operate with Class 9301 trailer cars (see below).
**UIC Numbers:** 8 039301 etc.

**Maximum Speed:** 70 km/h.

| 9301 | SE | 9305 | SE | 9307 | SE | 9308 | SE | 9310 | SE |
|------|----|------|----|------|----|------|----|------|----|
| 9303 | SE | 9306 | SE | | | | | | |

## CLASS 9301          Trailer Car

**Built:** 1954 by Allan.
**Normal Formation:** B.
**Length:** 19.510 m.
**Weight:**
**Note:** Operate with Class 9301 power cars (see above).
**UIC Numbers:** 28 29301 etc.

**Seats:**
**Wheel Arrangement:** 2-2.
**Maximum Speed:** 70 km/h.

| 9301 | SE | 9303 | SE | 9305 | SE | 9307 | SE | 9308 | SE |
|------|----|------|----|------|----|------|----|------|----|
| 9302 | SE | 9304 | SE | 9306 | SE | | | | |

## CLASS 9401          3-Car Unit

**Built:** 1966-68 by Djuro Djakovic. Acquired from Zajednica Jugoslovenskih Zeleznica (JZ) in 1980. Refurbished from Class 9701, former numbers not known.
**Normal Formation:** B + AB + B.
**Engine:** Volvo.
**Length:**
**Weight:**
**UIC Numbers:** 7 019401 + 7 019421 + 7 019411 etc.

**Seats:** 18F, 92S.
**Transmission:** Hydraulic. Voith.
**Wheel Arrangement:**
**Maximum Speed:** 60 km/h.

| 9401 | 9402 | 9403 | SE |
|------|------|------|----|

## CLASS 9631          2-Car Unit

**Built:** 1991 by SOREFAME.
**Normal Formation:**
**Engines:**
**Length:**
**Weight:**
**UIC Numbers:** 8 039631 + 5 009631 etc.

**Seats:** 100.
**Transmission:** Electric. ABB.
**Wheel Arrangement:**
**Maximum Speed:** 90 km/h.

| 9631 | PB | 9633 | PB | 9635 | PB | 9636 | PB | 9637 | PB |
|------|----|------|----|------|----|------|----|------|----|
| 9632 | PB | 9634 | PB | | | | | | |

## CLASS 9701          Driving Motor

**Built:** 1966-68 by Djuro Djakovic. Acquired from Zajednica Jugoslovenskih Zeleznica (JZ) in 1980.
**Normal Formation:**
**Engine:** Fiat of 137 kW.
**Length:**
**Weight:**
**Notes:** Some cars have been refurbished and re-engined with Volvo engines.
Six have been converted to Class 9401, but as the former identities of these six are not known, all cars purchased from JZ are listed below.
Allocated to MI and RE.
**UIC Numbers:** 6 019701 etc.

**Seats:**
**Transmission:** Mechanical. Fiat.
**Wheel Arrangement:** 2-B.
**Maximum Speed:** 60 km/h.

| 9701 | 9705 | 9709 | 9713 | 9717 |
|------|------|------|------|------|
| 9702 | 9706 | 9710 | 9714 | 9718 |
| 9703 | 9707 | 9711 | 9715 | 9719 |
| 9704 | 9708 | 9712 | 9716 | 9720 |

## CLASS 9721          Non Driving Motor

**Built:** 1966-68 by Djuro Djakovic. Acquired from Zajednica Jugoslovenskih Zeleznica (JZ) in 1980.
**Normal Formation:**

**Seats:**

**Engine:** Fiat of 137 kW.
**Length:**
**Weight:**
**Transmission:** Mechanical. Fiat.
**Wheel Arrangement:** 2-B.
**Maximum Speed:** 60 km/h.
**Notes:** Some cars have been refurbished and re-engined with Volvo engines.
Three have been converted to Class 9401, but as the former identities of these three are not known, all cars purchased from JZ are listed below.
Allocated to MI and RE. Some cars are stored out of use.
**UIC Numbers:** 6 019721 etc.

| | | | | |
|---|---|---|---|---|
| 9721 | 9725 | 9729 | 9733 | 9737 |
| 9722 | 9726 | 9730 | 9734 | 9738 |
| 9723 | 9727 | 9731 | 9735 | 9739 |
| 9724 | 9728 | 9732 | 9736 | 9740 |

# ESTORIL LINE ELECTRIC LOCOMOTIVES

**Electrification System:** 1500 V dc overhead.

## L301 Bo-Bo

**Built:** 1948 by North British.
**Electrical Equipment:** GEC.
**Weight:** 51.00 tonnes.
**Maximum Speed:**
L301 LC
**Continuous Rating:**
**Length:**
**Train Supply:** Not equipped.

## L302 Bo-Bo

**Built:** 1924 by AEG. Rebuilt 1948 by SOREFAME.
**Electrical Equipment:**
**Weight:**
**Maximum Speed:**
L302 LC
**Continuous Rating:**
**Length:**
**Train Supply:** Not equipped.

# ESTORIL LINE ELECTRIC MULTIPLE UNITS

**Electrification System:** 1500 V dc overhead.

## 101-111 3-Car Unit

**Built:** 1950 by Cravens.
**Normal Formation:** B + B + B.
**Electrical Equipment:** GEC.
**Motors:**
**Weight:**
**Seats:**
**Length:**
**Wheel Arrangement:**
**Maximum Speed:**

| | | | | | | | | | |
|---|---|---|---|---|---|---|---|---|---|
| 101 | LC | 104 | LC | 106 | LC | 108 | LC | 110 | LC |
| 102 | LC | 105 | LC | 107 | LC | 109 | LC | 111 | LC |
| 103 | LC | | | | | | | | |

## 112-124 3-Car Unit

**Built:** 1959-61 by SOREFAME.
**Normal Formation:** B + B + B.
**Electrical Equipment:** GEC.
**Motors:**
**Weight:** 105.40 tonnes.
**Seats:** 199S.
**Length:**
**Wheel Arrangement:**
**Maximum Speed:**

| | | | | | | | | | |
|---|---|---|---|---|---|---|---|---|---|
| 112 | LC | 115 | LC | 118 | LC | 121 | LC | 123 | LC |
| 113 | LC | 116 | LC | 119 | LC | 122 | LC | 124 | LC |
| 114 | LC | 117 | LC | 120 | LC | | | | |

## 201-214                                        4-Car Unit

**Built:** 1970 by SOREFAME.
**Normal Formation:** B + B + B + B.
**Electrical Equipment:** GEC.
**Motors:**
**Weight:**

**Seats:** 254S.
**Length:**
**Wheel Arrangement:**
**Maximum Speed:**

| | | | | | | | | | |
|---|---|---|---|---|---|---|---|---|---|
| 201 | LC | 204 | LC | 207 | LC | 210 | LC | 213 | LC |
| 202 | LC | 205 | LC | 208 | LC | 211 | LC | 214 | LC |
| 203 | LC | 206 | LC | 209 | LC | 212 | LC | | |

## 215-222                                        4-Car Unit

**Built:** 1979 by SOREFAME.
**Normal Formation:** B + B + B + B.
**Electrical Equipment:** GEC.
**Motors:**
**Weight:**

**Seats:**
**Length:**
**Wheel Arrangement:**
**Maximum Speed:**

| | | | | | | | | | |
|---|---|---|---|---|---|---|---|---|---|
| 215 | LC | 217 | LC | 219 | LC | 221 | LC | 222 | LC |
| 216 | LC | 218 | LC | 220 | LC | | | | |

# SPAIN

## VISAS

Not required by UK citizens for visits not exceeding 90 days. (NB: British subjects holding passports issued in Gibraltar do require visas). Full UK Passport necessary. or British Visitors Passport not valid after 31.09.95.

## BRITISH EMBASSY

**Address:** Calle de Fernando el Santo 16, 28010 Madrid. (Tel: (1) 319 0200; FAX: (1) 319 0423).
There are British Consulates in Algeciras, Alicante, Barcelona, Bilbao, Málaga, Santander, Seville, Tarragona and Vigo.

## CURRENCY & BANKING

Spanish Peseta (Pta). Notes are in denominations of Pta 10,000, 5000, 2000, 1000 and 500. Coins are in denominations of Pta 500, 200, 100, 50, 25, 10, 5, 2 and 1.
Banking hours are usually 0830-1630 Monday to Thursday, 0830-1400 Friday and 0830-1300 Saturday. In Summer (June to September) many banks only open 0830-1400 Monday to Friday.
Visa, Access/Mastercard, American Express and Diners Club are accepted at many outlets, as are Travellers Cheques and Eurocheques.

## OTHER USEFUL INFORMATION

**Language:** Spanish (Castilian), Catalan, Galican and Basque.
**Time:** GMT + 1 hour (GMT + 2 hours in summer).
**Electricity:** 220 V ac 50 Hz. Continental 2-pin plugs are standard.
**Usual Public Holidays:** New Year's Day, Epiphany, Maundy Thursday, Good Friday, Easter Monday (Barcelona only), May 1 (St. Joseph's Day), May 15 (St. Isidro's day - Madrid only), Corpus Christi, June 24 (King Juan Carlos' Saints Day), July 25 (St. James of Compostela Day), Assumption, October 12 (National Day), All Saints' Day, December 6 (Constitution Day), December 8 (Immaculate Conception Day - not Barcelona), Christmas Day, Boxing Day (Barcelona only).

## TOURIST INFORMATION

Contact Spanish National Tourist Office, Metro House, 57-58 St. James's Street, London, SW1A 1LD. (Tel: 0171-499-0901; FAX: 0171-629-4257).

# ACKNOWLEDGEMENT

Thanks are due to Philip Wormald for his help in compiling and checking this section.

# Red Nacional de los Ferrocarriles Españoles (RENFE)

**Gauges:** 1668 mm; 1435 mm; 1000 mm.
**Route Length:** 12111 km (1668mm); 471 Km (1435 mm); 19 km (1000 mm).
**Electrification System:** 3000 V dc overhead.

# NUMBERING SCHEME

The present numbering scheme dates from the early 1970s when a computerised system was introduced.
Virtually all vehicles built prior to the renumbering still carry their former numbers in addition to the computerised ones. Details of these are shown
in class headings to assist readers in identification.
Electric Locomotives are numbered in the 200 series of classes.
Diesel Locomotives are numbered in the 300 series of classes.
TALGO Locomotives are numbered in the 350 series of classes.
Electric Multiple Units are numbered in the 400 series of classes.
Diesel Multiple Units are numbered in the 500 series of classes.
A computer check digit appears after all numbers, but these are not shown in this book.

# DEPOTS

The system is divided into six zones, each with its own maintenance depots. The zone numbering scheme shown below and the depot codes used are the official scheme.

**MADRID (Zone 1)**
AC     Madrid Aravaca
AJ     Alcázar de San Juan
CN    Madrid Cerro Negro
FU    Madrid Fuencarral
MA   Madrid Atocha

**LEON (Zone 2)**
LE    León
OE   Orense
OV   Oviedo
SL    Salamanca

**SEVILLA (Zone 3)**
AM   Almería
GR   Granada
SP   Sevilla San Pablo
SJ    Sevilla Santa Justa

**VALENCIA (Zone 4)**
VF    Valencia Fuente de San Luis

**BARCELONA (Zone 5)**
BK   Barcelona Casa Antunez
BQ   Barcelona San Andres Condal
CC   Canfranc
RP   Ripoll
TA   Tarragona
ZD   Zaragoza-Delicias

**BILBAO (Zone 6)**
IR    Irún
ME   Miranda de Ebro
OG   Bilbao Olaveaga
ST   Santander

The AVE depot is at La Sagra. This is denoted in this book by the code LS.

# WORKSHOPS

The main RENFE workshops are located at Madrid Villaverde Bajo. Work is also carried out by various private contractors.

# RAILROVER INFORMATION

See the front of this book for details of Eurodomino tickets.

# SPAIN

## TIMETABLE

The "Guia RENFE I-Horarios" is a very elusive publication, but an Inter City train guide is more readily available for purchase at major stations. Readers are advised to be persistent in their enquiries at major stations in Spain.

## ELECTRIC LOCOMOTIVES

**Note:** All locomotives are 1668 mm gauge unless otherwise stated.

## CLASS 250.0                                                        C-C

**Built:** 1982-85 by Krauss Maffei/MTM/CAF.
**Electrical Equipment:** BBC.          **Continuous Rating:** 4600 kW.
**Weight:** 124.00 tonnes.               **Length:** 20.000 m.
**Maximum Speed:** 160 km/h.            **Train Supply:** Electric.

| | | | | | | | | | |
|---|---|---|---|---|---|---|---|---|---|
| 250.001 | VF | 250.008 | VF | 250.015 | VF | 250.022 | VF | 250.029 | VF |
| 250.002 | VF | 250.009 | VF | 250.016 | VF | 250.023 | VF | 250.030 | VF |
| 250.003 | VF | 250.010 | VF | 250.017 | VF | 250.024 | VF | 250.031 | VF |
| 250.004 | VF | 250.011 | VF | 250.018 | VF | 250.025 | VF | 250.032 | VF |
| 250.005 | VF | 250.012 | VF | 250.019 | VF | 250.026 | VF | 250.033 | VF |
| 250.006 | VF | 250.013 | VF | 250.020 | VF | 250.027 | VF | 250.034 | VF |
| 250.007 | VF | 250.014 | VF | 250.021 | VF | 250.028 | VF | 250.035 | VF |

## CLASS 250.6                                                        C-C

**Built:** 1986-88 by Krauss Maffei/CAF.
**Electrical Equipment:** BBC.          **Continuous Rating:** 4600 kW.
**Weight:** 130.00 tonnes.               **Length:** 20.000 m.
**Maximum Speed:** 160 km/h.            **Train Supply:** Electric.

| | | | | | | | | | |
|---|---|---|---|---|---|---|---|---|---|
| 250.601 | BK | 250.602 | BK | 250.603 | BK | 250.604 | BK | 250.605 | BK |

## CLASS 251                                                       B-B-B

**Built:** 1982-84 by Melco/CAF/Meinfesa.
**Electrical Equipment:** Melco.        **Continuous Rating:** 4650 kW.
**Weight:** 138.00 tonnes.               **Length:** 20.700 m.
**Maximum Speed:** 160 km/h.            **Train Supply:** Electric.

| | | | | | | | | | |
|---|---|---|---|---|---|---|---|---|---|
| 251.001 | OV | 251.007 | OV | 251.013 | OV | 251.019 | OV | 251.025 | OV |
| 251.002 | OV | 251.008 | OV | 251.014 | OV | 251.020 | OV | 251.026 | OV |
| 251.003 | OV | 251.009 | OV | 251.015 | OV | 251.021 | OV | 251.027 | OV |
| 251.004 | OV | 251.010 | OV | 251.016 | OV | 251.022 | OV | 251.028 | OV |
| 251.005 | OV | 251.011 | OV | 251.017 | OV | 251.023 | OV | 251.029 | OV |
| 251.006 | OV | 251.012 | OV | 251.018 | OV | 251.024 | OV | 251.030 | OV |

## CLASS 252                                                       Bo-Bo

**Built:** 1991-93 by Krauss Maffei/Henschel.
**Gauge:** 1435 mm.
**Supply System:** 25 kV ac 50 Hz or 3000 V dc overhead.
**Electrical Equipment:** Siemens.      **Continuous Rating:** 5600 kW.
**Weight:** 90.00 tonnes.                **Length:** 20.380 m.
**Maximum Speed:** 220 km/h.            **Train Supply:** Electric.

| | | | | | | | | | |
|---|---|---|---|---|---|---|---|---|---|
| 252.001 | LS | 252.004 | LS | 252.007 | LS | 252.010 | LS | 252.013 | LS |
| 252.002 | LS | 252.005 | LS | 252.008 | LS | 252.011 | LS | 252.014 | LS |
| 252.003 | LS | 252.006 | LS | 252.009 | LS | 252.012 | LS | 252.015 | LS |

## CLASS 252                                                       Bo-Bo

**Built:** 1993-1995 by Meinfesa/CAF.
**Supply System:** 3000 V dc overhead (* 25 kV ac 50 Hz or 3000 V dc overhead).

**Electrical Equipment:** Siemens.
**Weight:** 86.00 (* 90.00) tonnes.
**Maximum Speed:** 160 km/h.

**Continuous Rating:** 5600 kW.
**Length:** 20.380 m.
**Train Supply:** Electric.

| | | | | | | | | | |
|---|---|---|---|---|---|---|---|---|---|
| 252.016* | BK | 252.028* | BK | 252.040 | BK | 252.052 | BK | 252.064 | BK |
| 252.017* | BK | 252.029* | BK | 252.041 | BK | 252.053 | BK | 252.065 | BK |
| 252.018* | BK | 252.030* | BK | 252.042 | BK | 252.054 | BK | 252.066 | BK |
| 252.019* | BK | 252.031* | BK | 252.043 | BK | 252.055 | BK | 252.067 | BK |
| 252.020* | BK | 252.032 | BK | 252.044 | BK | 252.056 | BK | 252.068 | BK |
| 252.021* | BK | 252.033 | BK | 252.045 | BK | 252.057 | BK | 252.069 | BK |
| 252.022* | BK | 252.034 | BK | 252.046 | BK | 252.058 | BK | 252.070 | BK |
| 252.023* | BK | 252.035 | BK | 252.047 | BK | 252.059 | BK | 252.071 | BK |
| 252.024* | BK | 252.036 | BK | 252.048 | BK | 252.060 | BK | 252.072 | BK |
| 252.025* | BK | 252.037 | BK | 252.049 | BK | 252.061 | BK | 252.073 | BK |
| 252.026* | BK | 252.038 | BK | 252.050 | BK | 252.062 | BK | 252.074 | BK |
| 252.027* | BK | 252.039 | BK | 252.051 | BK | 252.063 | BK | 252.075 | BK |

# CLASS 269.0        B-B

**Built:** 1973-78 by Atiensa/CAF/Melco/Westinghouse.
**Electrical Equipment:** Melco.
**Weight:** 88.00 tonnes.
**Maximum Speed:** 140 km/h.

**Continuous Rating:** 3100 kW.
**Length:** 17.270 m.
**Train Supply:** Electric.

| | | | | | | | | | |
|---|---|---|---|---|---|---|---|---|---|
| 269.001 | LE | 269.023 | ST | 269.044 | ST | 269.066 | BK | 269.088 | ME |
| 269.002 | LE | 269.024 | ST | 269.045 | ST | 269.067 | BK | 269.089 | BK |
| 269.003 | LE | 269.025 | ST | 269.046 | ST | 269.068 | BK | 269.090 | ME |
| 269.004 | LE | 269.026 | ST | 269.048 | ST | 269.069 | BK | 269.091 | ME |
| 269.005 | LE | 269.027 | ST | 269.049 | ST | 269.070 | BK | 269.092 | ME |
| 269.006 | BK | 269.028 | ST | 269.050 | ST | 269.071 | BK | 269.093 | AJ |
| 269.007 | BK | 269.029 | ST | 269.051 | ST | 269.072 | BK | 269.095 | ME |
| 269.008 | LE | 269.030 | ST | 269.052 | ST | 269.073 | BK | 269.096 | ME |
| 269.009 | LE | 269.031 | ST | 269.053 | AJ | 269.074 | BK | 269.097 | AJ |
| 269.010 | AJ | 269.032 | ST | 269.054 | AJ | 269.075 | BK | 269.098 | ME |
| 269.011 | LE | 269.033 | ST | 269.055 | AJ | 269.076 | BK | 269.099 | ME |
| 269.012 | AJ | 269.034 | ST | 269.056 | AJ | 269.077 | BK | 269.100 | ME |
| 269.013 | LE | 269.035 | ST | 269.057 | BK | 269.078 | LE | 269.101 | AJ |
| 269.014 | LE | 269.036 | ST | 269.058 | AJ | 269.079 | LE | 269.102 | AJ |
| 269.015 | LE | 269.037 | ST | 269.059 | BK | 269.081 | LE | 269.103 | AJ |
| 269.016 | LE | 269.038 | ST | 269.060 | BK | 269.082 | BK | 269.104 | AJ |
| 269.018 | LE | 269.039 | ST | 269.061 | BK | 269.083 | BK | 269.105 | AJ |
| 269.019 | ST | 269.040 | ST | 269.062 | BK | 269.084 | BK | 269.106 | AJ |
| 269.020 | ST | 269.041 | ST | 269.063 | BK | 269.085 | BK | 269.107 | AJ |
| 269.021 | ST | 269.042 | ST | 269.064 | AJ | 269.086 | BK | 269.108 | AJ |
| 269.022 | ST | 269.043 | ST | 269.065 | BK | 269.087 | BK | | |

# CLASS 269.2        B-B

**Built:** 1983-85 by CAF/Meinfesa/Atiensa/Melco/Westinghouse/GEE.
**Electrical Equipment:** Melco.
**Weight:** 88.00 tonnes.
**Maximum Speed:** 160 (* 140) km/h.

**Continuous Rating:** 3100 kW.
**Length:** 17.270 m.
**Train Supply:** Electric.

| | | | | | | | | | |
|---|---|---|---|---|---|---|---|---|---|
| 269.201 | SJ | 269.213 | SJ | 269.225 | SJ | 269.237 | SJ | 269.249 | SJ |
| 269.202 | SJ | 269.214 | SJ | 269.226* | AJ | 269.238 | SJ | 269.250 | SJ |
| 269.203 | SJ | 269.215 | SJ | 269 227 | SJ | 269.239 | SJ | 269.251 | SJ |
| 269.204 | SJ | 269.216 | SJ | 269.228 | LE | 269.240 | SJ | 269.252* | SJ |
| 269.205 | SJ | 269.217 | SJ | 269.229 | LE | 269.241 | SJ | 269.253 | SJ |
| 269.206 | SJ | 269.218 | SJ | 269.230 | SJ | 269.242 | ST | 269.254 | SJ |
| 269.207 | SJ | 269.219 | SJ | 269.231* | SJ | 269.243 | ST | 269.255 | SJ |
| 269.208 | SJ | 269.220 | SJ | 269.232 | SJ | 269.244 | SJ | 269.256 | SJ |
| 269.209 | SJ | 269.221 | SJ | 269.233 | SJ | 269.245 | SJ | 269.257 | SJ |
| 269.210 | SJ | 269.222 | ST | 269.234 | ST | 269.246 | SJ | 269.258 | AJ |
| 269.211 | SJ | 269.223 | SJ | 269.235* | AJ | 269.247* | BK | 269.259 | AJ |
| 269.212 | SJ | 269.224 | SJ | 269.236 | SJ | 269.248 | SJ | 269.260 | AJ |

# SPAIN

| | | | | | | | | | |
|---|---|---|---|---|---|---|---|---|---|
| 269.261 | LE | 269.276 | LE | 269.290 | BK | 269.304 | ST | 269.318 | ME |
| 269.262 | LE | 269.277 | LE | 269.291 | BK | 269.305 | ST | 269.319 | ME |
| 269.263 | AJ | 269.278 | BK | 269.292* | BK | 269.306 | ST | 269.320 | ME |
| 269.264 | AJ | 269.279 | LE | 269.293 | BK | 269.307 | ST | 269.321 | ME |
| 269.265 | SJ | 269.280 | BK | 269.294 | BK | 269.308 | ST | 269.322* | SJ |
| 269.266 | ME | 269.281 | BK | 269.295 | BK | 269.309 | ST | 269.323 | ME |
| 269.267 | ME | 269.282 | LE | 269.296* | FU | 269.310* | FU | 269.324 | ME |
| 269.268 | BK | 269.283 | LE | 269.297* | FU | 269.311 | ST | 269.325 | ME |
| 269.269* | BK | 269.284 | LE | 269.298 | BK | 269.312 | ST | 269.326 | ME |
| 269.270 | BK | 269.285 | BK | 269.299* | FU | 269.313* | FU | 269.327 | ME |
| 269.271 | BK | 269.286 | BK | 269.300 | BK | 269.314 | LE | 269.328 | ME |
| 269.272 | BK | 269.287 | BK | 269.301 | AJ | 269.315* | BK | 269.329 | ME |
| 269.273 | LE | 269.288 | LE | 269.302 | AJ | 269.316 | BK | 269.330 | ME |
| 269.274 | LE | 269.289 | LE | 269.303 | AJ | 269.317 | LE | 269.331* | SJ |
| 269.275 | BK | | | | | | | | |

## CLASS 269.5                                                                 B-B

**Built:** 1974-79 by CAF/Melco/Westinghouse/GEE.
**Electrical Equipment:** Melco.          **Continuous Rating:** 3100 kW.
**Weight:** 88.00 tonnes.                 **Length:** 17.270 m.
**Maximum Speed:** 160 km/h.              **Train Supply:** Electric.

| | | | | | | | | | |
|---|---|---|---|---|---|---|---|---|---|
| 269.501 | LE | 269.506 | LE | 269.511 | ME | 269.515 | ME | 269.519 | ME |
| 269.502 | LE | 269.507 | LE | 269.512 | ME | 269.516 | ME | 269.520 | ME |
| 269.503 | LE | 269.508 | LE | 269.513 | ME | 269.517 | ME | 269.521 | ME |
| 269.504 | LE | 269.509 | LE | 269.514 | ME | 269.518 | ME | 269.522 | ME |
| 269 505 | LE | 269 510 | LE | | | | | | |

## CLASS 269.6                                                                 B-B

**Built:** 1980-81 by Melco/CAF/Westinghouse. Rebuilt 1988 by CAF.
**Electrical Equipment:** Melco.          **Continuous Rating:** 3100 kW.
**Weight:** 80.00 tonnes.                 **Length:** 17.270 m.
**Maximum Speed:** 200 km/h.              **Train Supply:** Electric.

| | | | | | | | |
|---|---|---|---|---|---|---|---|
| 269.601 | FU | 269.602 | FU | 269.603 | FU | 269.604 | FU |

## CLASS 276                                                                 Co-Co

**Built:** 1956-63 by Alsthom/Macosa/CAF/Euskalduna.
**Electrical Equipment:** Alsthom.        **Continuous Rating:** 2200 kW.
**Weight:** 120.00 tonnes.                **Length:** 18.932 m.
**Maximum Speed:** 110 km/h.              **Train Supply:** Electric.
**Former Numbers:** 7601-99 (276.001-099 in sequence), 8601-37 (276.101-137 in sequence).

| | | | | | | | | | |
|---|---|---|---|---|---|---|---|---|---|
| 276.002 | ZD | 276.031 | ZD | 276.052 | ZD | 276.086 | ZD | 276.111 | AJ |
| 276.005 | AJ | 276.033 | ZD | 276.059 | ZD | 276.097 | ZD | 276.116 | AJ |
| 276.007 | AJ | 276.036 | AJ | 276.066 | ZD | 276.099 | ZD | 276.123 | AJ |
| 276.013 | ZD | 276.037 | AJ | 276.068 | ZD | 276.101 | AJ | 276.125 | AJ |
| 276.014 | ZD | 276.043 | ZD | 276.073 | ZD | 276.102 | AJ | 276.126 | AJ |
| 276.019 | ZD | 276.046 | ZD | 276.080 | ZD | 276.107 | AJ | 276.130 | AJ |
| 276.021 | AJ | 276.048 | ZD | 276.084 | ZD | 276.109 | AJ | | |

## CLASS 276.2                                                              Co-Co

**Built:** 1956-63 by Alsthom/Macosa/CAF/Euskalduna. Rebuilt 1993.
**Electrical Equipment:** Alsthom.        **Continuous Rating:** 2200 kW.
**Weight:** 120.00 tonnes.                **Length:** 18.932 m.
**Maximum Speed:** 110 km/h.              **Train Supply:** Electric.

| | | | | | |
|---|---|---|---|---|---|
| 276.201 (276.110) | VF | 276.205 (276.028) | VF | 276.208 (276.131) | VF |
| 276.202 (276.091) | VF | 276.206 (276.063) | VF | 276.209 (276.079) | VF |
| 276.203 (276.060) | VF | 276.207 (276.045) | VF | 276.210 (276.082) | VF |
| 276 204 (276 133) | VF | | | | |

## CLASS 279      Bo-Bo

**Built:** 1967 by Melco/CAF/Cenemesa.
**Supply System:** 3000 v dc or 1500 V dc overhead.
**Electrical Equipment:** Melco.      **Continuous Rating:** 2700 kW.
**Weight:** 80.00 tonnes.      **Length:** 17.270 m.
**Maximum Speed:** 130 km/h.      **Train Supply:** Electric.
**Former Numbers:** 7901-16 in sequence.

| | | | | |
|---|---|---|---|---|
| 279.001 ME | 279.005 ME | 279.008 ME | 279.011 ME | 279.014 ME |
| 279.002 ME | 279.006 ME | 279.009 ME | 279.012 ME | 279.015 ME |
| 279.003 ME | 279.007 ME | 279.010 ME | 279.013 ME | 279.016 ME |
| 279 004 ME | | | | |

## CLASS 289      Bo-Bo

**Built:** 1969-72 by Melco/CAF/Cenemesa.
**Supply System:** 3000 V dc or 1500 V dc overhead.
**Electrical Equipment:** Melco.      **Continuous Rating:** 3100 kW.
**Weight:** 84.00 tonnes.      **Length:** 17.270 m.
**Maximum Speed:** 130 km/h.      **Train Supply:** Electric.
**Former Numbers:** 8901-40 in sequence.

| | | | | |
|---|---|---|---|---|
| 289.001 ME | 289.009 ME | 289.017 ME | 289.025 AM | 289.033 ME |
| 289.002 ME | 289.010 ME | 289.018 ME | 289.026 ME | 289.034 ME |
| 289.003 ME | 289.011 AM | 289.019 ME | 289.027 ME | 289.035 ME |
| 289.004 AM | 289.012 ME | 289.020 ME | 289.028 AM | 289.036 ME |
| 289.005 ME | 289.013 ME | 289.021 AM | 289.029 ME | 289.037 ME |
| 289.006 ME | 289.014 ME | 289.022 ME | 289.030 ME | 289.038 ME |
| 289.007 AM | 289.015 ME | 289.023 ME | 289.031 ME | 289.039 ME |
| 289.008 ME | 289.016 ME | 289.024 ME | 289.032 ME | 289.040 ME |

# DIESEL LOCOMOTIVES

## CLASS 303      C

**Built:** 1961-65 by MTM/Babcock/Sulzer/BBC.
**Engine:** Sulzer 6LD22A of 200 kW at 870 rpm.
**Transmission:** Electric. BBC.
**Weight:** 48.00 tonnes.      **Length:** 9.800 m.
**Maximum Speed:** 45 km/h.      **Train Supply:** Not equipped.
**Former Numbers:** 10301-400, 11301-400, 12301-302 (303.001-202) in sequence.

| | | | | |
|---|---|---|---|---|
| 303.004 MA | 303.044 ME | 303.132 IR | 303.181 IR | 303.201 MA |
| 303.017 MA | 303.049 ST | 303.139 MA | 303.187 ME | 303.202 MA |
| 303.020 MA | 303.096 SJ | 303.172 IR | 303.188 ME | |

## CLASS 304      C

**Built:** 1966-77 by MTM/Sulzer/BBC.
**Engine:** Sulzer 6LD22B of 295 kW at 950 rpm.
**Transmission:** Electric. BBC.
**Weight:** 48.00 tonnes.      **Length:** 9.800 m.
**Maximum Speed:** 45 km/h.      **Train Supply:** Not equipped.
**Former Numbers:** 10401-63 (304.001-063) in sequence, not all carried.

| | | | | |
|---|---|---|---|---|
| 304.010 BK | 304.013 BK | 304.030 BK | 304.037 BK | 304.045 BK |
| 304.011 BK | 304.018 BK | 304.035 BK | 304.040 BK | 304.063 BK |
| 304.012 BK | | | | |

## CLASS 307      Bo-Bo

**Built:** 1962-63 by B & L/MTM.

# SPAIN

**Engine:** Sulzer LDA22E of 400 kW at 950 rpm.
**Transmission:** Electric. B & L.
**Weight:** 66.00 tonnes. **Length:** 14.680 m.
**Maximum Speed:** 80 km/h. **Train Supply:** Not equipped.
**Former Numbers:** 10701-10 (307.001-010) in sequence.

| | | | | |
|---|---|---|---|---|
| 307.002 VF | 307.004 OV | 307.008 OV | 307.009 OV | 307.010 OV |
| 307.003 OV | 307.007 OV | | | |

# CLASS 308 Bo-Bo

**Built:** 1966-69 by General Electric/Babcock.
**Engine:** Caterpillar D-398-12 of 520 kW at 1300 rpm.
**Transmission:** Electric. General Electric.
**Weight:** 64.00 tonnes. **Length:** 12.935 m.
**Maximum Speed:** 114 km/h. **Train Supply:** Not equipped.
**Former Numbers:** 10801-41 in sequence.

| | | | | | | | | | |
|---|---|---|---|---|---|---|---|---|---|
| 308.001 | MA | 308.010 | VF | 308.018 | ME | 308.026 | MA | 308.034 | MA |
| 308.002 | ME | 308.011 | VF | 308.019 | VF | 308.027 | TA | 308.035 | MA |
| 308.003 | MA | 308.012 | VF | 308.020 | TA | 308.028 | MA | 308.036 | MA |
| 308.004 | MA | 308.013 | TA | 308.021 | ME | 308.029 | ME | 308.037 | OE |
| 308.005 | ME | 308.014 | OE | 308.022 | BK | 308.030 | VF | 308.038 | MA |
| 308.006 | VF | 308.015 | MA | 308.023 | OE | 308.031 | MA | 308.039 | ME |
| 308.007 | VF | 308.016 | BK | 308.024 | TA | 308.032 | ME | 308.040 | MA |
| 308.008 | BK | 308.017 | TA | 308.025 | VF | 308.033 | MA | 308.041 | ME |
| 308 009 | MA | | | | | | | | |

# CLASS 309 C

**Built:** 1987 by MTM/Bazán.
**Engine:** MTU 6V396TC13 of 385 kW at 1900 rpm.
**Transmission:** Hydraulic. Voith.
**Weight:** 54.00 tonnes. **Length:** 9.960 m.
**Maximum Speed:** 50 km/h. **Train Supply:** Not equipped.

| | | | | | |
|---|---|---|---|---|---|
| 309.001 ME | 309.005 ST | 309.009 ST | 309.013 IR | 309.017 ME |
| 309.002 OG | 309.006 ST | 309.010 ST | 309.014 IR | 309.018 IR |
| 309.003 ST | 309.007 FG | 309.011 IR | 309.015 OG | 309.019 ME |
| 309.004 IR | 309.008 OG | 309.012 ME | 309.016 OG | 309.020 IR |

# CLASS 310 Bo-Bo

**Built:** 1989-91 by Meinfesa/Westinghouse.
**Engine:** GM 8-645E of 600 kW at 900 rpm.
**Transmission:** Electric. GM.
**Weight:** 78.00 tonnes. **Length:** 12.550 m.
**Maximum Speed:** 115 km/h. **Train Supply:** Not equipped.

| | | | | |
|---|---|---|---|---|
| 310.001 VF | 310.013 SL | 310.025 VF | 310.037 SJ | 310.049 VF |
| 310.002 VF | 310.014 AJ | 310.026 VF | 310.038 SJ | 310.050 SL |
| 310.003 VF | 310.015 AJ | 310.027 SJ | 310.039 MA | 310.051 SL |
| 310.004 VF | 310.016 ZD | 310.028 SL | 310.040 MA | 310.052 SL |
| 310.005 VF | 310.017 ZD | 310.029 VF | 310.041 MA | 310.053 ZD |
| 310.006 SJ | 310.018 ZD | 310.030 VF | 310.042 AJ | 310.054 ZD |
| 310.007 SJ | 310.019 SJ | 310.031 ZD | 310.043 VF | 310.055 ZD |
| 310.008 SJ | 310.020 SJ | 310.032 MA | 310.044 VF | 310.056 AJ |
| 310.009 SJ | 310.021 SJ | 310.033 VF | 310.045 VF | 310.057 ZD |
| 310.010 SJ | 310.022 VF | 310.034 SJ | 310.046 VF | 310.058 SL |
| 310.011 SJ | 310.023 VF | 310.035 SJ | 310.047 SJ | 310.059 SL |
| 310.012 SL | 310.024 VF | 310.036 SJ | 310.048 SL | 310.060 MA |

# CLASS 311 Bo-Bo

**Built:** 1985-91 by MTM/Atiensa/Babcock/Alsthom/Siemens.
**Engine:** Bazán MTU 8V396TC13 of 504 kW at 1800 rpm.

**Transmission:** Electric. Siemens.
**Weight:** 80.00 tonnes.  **Length:** 14.200 m.
**Maximum Speed:** 90 km/h.  **Train Supply:** Not equipped.

| | | | | | | | | | |
|---|---|---|---|---|---|---|---|---|---|
| 311.001 | BK | 311.113 | BK | 311.125 | FU | 311.137 | BK | 311.149 | OV |
| 311.101 | OG | 311.114 | FU | 311.126 | BK | 311.138 | FU | 311.150 | OE |
| 311.102 | BK | 311.115 | BK | 311.127 | FU | 311.139 | FU | 311.151 | OE |
| 311.103 | BK | 311.116 | BK | 311.128 | FU | 311.140 | LE | 311.152 | OE |
| 311.104 | BK | 311.117 | BK | 311.129 | BK | 311.141 | OV | 311.153 | OE |
| 311.105 | BK | 311.118 | FU | 311.130 | BK | 311.142 | BK | 311.154 | OV |
| 311.106 | FU | 311.119 | FU | 311.131 | BK | 311.143 | BK | 311.155 | OV |
| 311.107 | FU | 311.120 | OE | 311.132 | LE | 311.144 | LE | 311.156 | OE |
| 311.108 | FU | 311.121 | TA | 311.133 | LE | 311.145 | LE | 311.157 | OE |
| 311.109 | BK | 311.122 | TA | 311.134 | OG | 311.146 | BK | 311.158 | OV |
| 311.110 | OE | 311.123 | TA | 311.135 | FU | 311.147 | BK | 311.159 | OE |
| 311.111 | FU | 311.124 | FU | 311.136 | BK | 311.148 | OV | 311.160 | BK |
| 311.112 | BK | | | | | | | | |

## CLASS 313 — Co-Co

**Built:** 1965-67 by Euskalduna/Alco.
**Engine:** Alco 251D-6 TC13 of 743 kW at 1100 rpm.
**Transmission:** Electric. General Electric.
**Weight:** 83.92 tonnes.  **Length:** 16.237 m.
**Maximum Speed:** 120 km/h.  **Train Supply:** Not equipped.
**Former Numbers:** 1301-50 (313.001-050) in sequence.

| | | | | | | | | | |
|---|---|---|---|---|---|---|---|---|---|
| 313.006 | GR | 313.018 | GR | 313.023 | GR | 313.036 | GR | 313.044 | GR |
| 313.010 | GR | 313.021 | GR | 313.026 | GR | 313.042 | GR | 313.046 | GR |
| 313.016 | GR | | | | | | | | |

## CLASS 319.2 — Co-Co

**Gauge:** 1668 († 1435) mm.
**Built:** 1965-72 by Macosa/GM. Rebuilt 1984 onwards by Meinfesa/MTM.
**Engine:** GM 16-567C of 1154 kW at 835 rpm.
**Transmission:** Electric. GM.
**Weight:** 110.00 (*† 112.00) tonnes.  **Length:** 19.500 m.
**Maximum Speed:** 120 km/h.  **Train Supply:** Not equipped.

| | | | | | | | | | |
|---|---|---|---|---|---|---|---|---|---|
| 319.201 | ZD | 319.213 | ZD | 319.225* | GR | 319.237* | VF | 319.248† | LS |
| 319.202 | ZD | 319.214 | ZD | 319.226* | GR | 319.238* | VF | 319 249 | |
| 319.203 | ZD | 319.215 | ZD | 319.227* | GR | 319.239* | VF | 319 250 | |
| 319.204 | ZD | 319.216 | ZD | 319.228* | GR | 319.240* | VF | 319 251 | |
| 319.205 | ZD | 319.217 | ZD | 319.229* | VF | 319.241† | LS | 319 252 | |
| 319.206 | ZD | 319.218 | ZD | 319.230* | VF | 319.242† | LS | 319 253 | |
| 319.207 | ZD | 319.219 | ZD | 319.231* | VF | 319.243† | LS | 319 254 | |
| 319.208 | ZD | 319.220 | ZD | 319.232* | VF | 319.244† | LS | 319 255 | |
| 319.209 | ZD | 319.221* | GR | 319.233* | VF | 319.245† | LS | 319 256 | |
| 319.210 | ZD | 319.222* | GR | 319.234* | VF | 319.246† | LS | 319 257 | |
| 319.211 | ZD | 319.223* | GR | 319.235* | VF | 319.247† | LS | 319 258 | |
| 319.212 | ZD | 319.224* | GR | 319.236* | VF | | | | |

## CLASS 319.3 — Co-Co

**Built:** 1965-72 by Macosa/GM. Rebuilt 1991-92 by Meinfesa.
**Engine:** GM 16-567C of 1490 kW at 900 rpm.
**Transmission:** Electric. GM.
**Weight:** 117.00 tonnes.  **Length:** 19.500 m.
**Maximum Speed:** 120 km/h.  **Train Supply:** Electric.

| | | | | | | | | | |
|---|---|---|---|---|---|---|---|---|---|
| 319.301 | GR | 319.306 | GR | 319.311 | GR | 319.316 | SL | 319.321 | OE |
| 319.302 | GR | 319.307 | GR | 319.312 | GR | 319.317 | OE | 319.322 | OE |
| 319.303 | GR | 319.308 | GR | 319.313 | SL | 319.318 | OE | 319.323 | OE |
| 319.304 | GR | 319.309 | GR | 319.314 | SL | 319.319. | OE | 319.324 | OE |
| 319.305 | GR | 319.310 | GR | 319.315 | SL | 319.320 | OE | 319.325 | OE |

# SPAIN

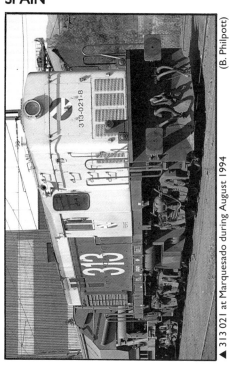

▲ 313 021 at Marquesado during August 1994 (B. Philpott)

▼ 333 079 at Medina del Campo with an Irún - Salamanca train on 14.04.95 (P. Wormald)

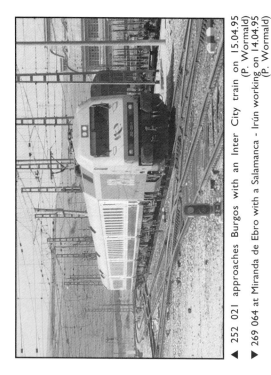

▲ 252 021 approaches Burgos with an Inter City train on 15.04.95 (P. Wormald)

▼ 269 064 at Miranda de Ebro with a Salamanca - Irún working on 14.04.95 (P. Wormald)

| 319.326 | OE | 319.329 | OE | 319.332 | GR | 319.335 | MA | 319.338 | MA |
| 319.327 | ZD | 319.330 | OE | 319.333 | GR | 319.336 | MA | 319.339 | MA |
| 319.328 | ZD | 319.331 | GR | 319.334 | MA | 319.337 | MA | 319.340 | MA |

# CLASS 319.4     Co-Co

**Built:** 1965-72 by Macosa/GM. Rebuilt 1992-93 by Meinfesa.
**Engine:** GM 16-567C of 1490 kw at 900 rpm.
**Transmission:** Electric. GM.
**Weight:** 116.00 tonnes.    **Length:** 19.500 m.
**Maximum Speed:** 120 km/h.    **Train Supply:** Not equipped.

| 319.401 | ZD | 319.403 | ZD | 319.405 | SJ | 319.407 | ZD | 319.409 | ZD |
| 319.402 | ZD | 319.404 | ZD | 319.406 | SJ | 319.408 | SJ | 319.410 | SJ |

# CLASS 321     Co-Co

**Built:** 1964-67 by Alco/CAF/SECN/Euskalduna.
**Engine:** Alco 251C-12 of 1250 kW at 1025 rpm.
**Transmission:** Electric. General Electric.
**Weight:** 111.00 tonnes.    **Length:** 18.567 m.
**Maximum Speed:** 120 km/h.    **Train Supply:** Not equipped.
**Former Numbers:** 2101-80 (321.001-080) in sequence.

| 321.002 | SJ | 321.019 | SJ | 321.035 | OE | 321.052 | OR | 321.067 | MA |
| 321.003 | SJ | 321.020 | SJ | 321.036 | OE | 321.053 | OR | 321.068 | MA |
| 321.004 | OE | 321.021 | SJ | 321.038 | SJ | 321.054 | OE | 321.069 | MA |
| 321.005 | GR | 321.022 | MA | 321.039 | SJ | 321.055 | GR | 321.070 | MA |
| 321.007 | SJ | 321.023 | SJ | 321.041 | SJ | 321.057 | OE | 321.071 | GR |
| 321.008 | SJ | 321.024 | SJ | 321.042 | OE | 321.058 | OE | 321.073 | MA |
| 321.009 | SJ | 321.026 | SJ | 321.044 | OE | 321.059 | SJ | 321.075 | GR |
| 321.010 | SJ | 321.027 | SJ | 321.046 | OE | 321.060 | OE | 321.077 | GR |
| 321.011 | SJ | 321.028 | MA | 321.048 | OE | 321.061 | MA | 321.078 | MA |
| 321.012 | SJ | 321.030 | OE | 321.050 | SJ | 321.063 | MA | 321.079 | MA |
| 321.014 | MA | 321.032 | OE | 321.051 | OE | 321.065 | MA | 321.080 | MA |
| 321.015 | SJ | 321.034 | SJ | | | | | | |

# CLASS 333     Co-Co

**Built:** 1974-76 by Macosa.
**Engine:** GM 16-645E3 of 1875 kW at 900 rpm.
**Transmission:** Electric. GM.
**Weight:** 120.00 tonnes.    **Length:** 20.700 m.
**Maximum Speed:** 146 km/h.    **Train Supply:** Electric.

| 333.001 | MA | 333.020 | MA | 333.039 | VF | 333.058 | VF | 333.076 | SL |
| 333.002 | VF | 333.021 | SJ | 333.040 | MA | 333.059 | MA | 333.077 | SJ |
| 333.003 | MA | 333.022 | SJ | 333.042 | MA | 333.060 | MA | 333.078 | SL |
| 333.004 | MA | 333.023 | VF | 333.043 | MA | 333.061 | MA | 333.079 | SL |
| 333.005 | MA | 333.024 | VF | 333.044 | MA | 333.062 | SL | 333.080 | SL |
| 333.006 | MA | 333.025 | VF | 333.045 | MA | 333.063 | SL | 333.081 | VF |
| 333.007 | MA | 333.026 | VF | 333.046 | MA | 333.064 | SL | 333.082 | SL |
| 333.008 | MA | 333.027 | SJ | 333.047 | MA | 333.065 | SL | 333.083 | SL |
| 333.009 | MA | 333.028 | SJ | 333.048 | MA | 333.066 | SL | 333.084 | SL |
| 333.010 | MA | 333.029 | SJ | 333.049 | MA | 333.067 | SL | 333.085 | SL |
| 333.011 | MA | 333.030 | SJ | 333.050 | MA | 333.068 | SL | 333.086 | OE |
| 333.012 | MA | 333.031 | VF | 333.051 | VF | 333.069 | SL | 333.087 | OE |
| 333.013 | SJ | 333.032 | SJ | 333.052 | VF | 333.070 | SJ | 333.088 | OE |
| 333.014 | MA | 333.033 | VF | 333.053 | VF | 333.071 | SL | 333.089 | OE |
| 333.015 | VF | 333.034 | VF | 333.054 | VF | 333.072 | SL | 333.090 | OE |
| 333.016 | MA | 333.035 | VF | 333.055 | SJ | 333.073 | SL | 333.091 | OE |
| 333.017 | MA | 333.036 | VF | 333.056 | VF | 333.074 | SL | 333.092 | SJ |
| 333.018 | MA | 333.037 | VF | 333.057 | VF | 333.075 | SL | 333.093 | OE |
| 333.019 | MA | 333.038 | VF | | | | | | |

# SPAIN

## CLASS 333.2 Co-Co

**Built:** 1974-76 by Macosa. Refurbished 1994 onwards by RENFE, Vilaverde Works.
**Engine:** GM 16-645E3 of 1875 kW at 900 rpm.
**Transmission:** Electric. GM.
**Weight:** 120.00 tonnes. **Length:** 20.700 m.
**Maximum Speed:** 160 km/h. **Train Supply:** Electric.

333.201 (333.041)　　MA　　333.203 (333.0　)
333.202 (333.0　)

## CLASS 352 (TALGO) B-B

**Built:** 1964-65 by Krauss-Maffei/Babcock.
**Engines:** Two Maybach MD650/1-12 of 550 kW each at 1500 rpm.
**Transmission:** Hydraulic. Mekydro K104u.
**Weight:** 74.00 tonnes. **Length:** 17.450 m.
**Maximum Speed:** 140 km/h. **Train Supply:** Electric.
**Former Numbers:** 2001T-010T (352.001-010) in sequence.

| | | | |
|---|---|---|---|
| 352.001 AC | *Virgen del Rosario* | 352.006 AC | *Virgen Santa Maria* |
| 352.002 AC | *Virgen Peregrina* | 352.007 AC | *Virgen de la Almudena* |
| 352.003 AC | *Virgen del Perpetuo Socorro* | 352.008 AC | *Virgen de la Soledad* |
| 352.004 AC | *Virgen del Camino* | 352.009 AC | *Virgen de Gracia* |
| 352.005 AC | *Virgen del Carmen* | 352.010 AC | *Virgen de los Reyes* |

## CLASS 353 (TALGO) B-B

**Built:** 1968-69 by Krauss-Maffei.
**Engines:** Two Maybach MD655Z-12 of 825 kW each at 1600 rpm.
**Transmission:** Hydraulic. Mekydro K184u.
**Weight:** 88.00 tonnes. **Length:** 19.000 m.
**Maximum Speed:** 180 km/h. **Train Supply:** Electric.
**Former Numbers:** 3001T-05T (353.001-005) in sequence.

| | | | |
|---|---|---|---|
| 353.001 AC | *Virgen de Lourdes* | 353.005 AC | *Virgen Bien Aparecida* |
| 353.002 AC | *Virgen de la Fatima* | | |

## CLASS 354 (TALGO) B-B

**Built:** 1983-84 by Krauss-Maffei.
**Engines:** Two MTU 16V396TD13 of 1125 kW each at 1900 rpm.
**Transmission:** Hydraulic. Voith L520rzU2.
**Weight:** 80.00 tonnes. **Length:** 19.920 m.
**Maximum Speed:** 200 km/h. **Train Supply:** Not equipped.

| | | | |
|---|---|---|---|
| 354.001 AC | *Virgen de Covadonga* | 354.005 AC | *Virgen del Pilar* |
| 354.002 AC | *Virgen de la Macarena* | 354.006 AC | *Virgen de Aránzazu* |
| 354.003 AC | *Virgen de la Encarnación* | 354.007 AC | *Virgen de Begoña* |
| 354.004 AC | *Virgen de Guadalupe* | 354.008 AC | *Virgen de Montserrat* |

# ELECTRIC MULTIPLE UNITS

**Note:** For EMU vehicles, only one number is quoted where all cars within a set carry the same number but with a different prefix. Individual car numbers are quoted where these vary between cars within a unit.

## AVE 10-Car Unit

**Built:** 1991-95 by Alsthom/MTM/Atiensa.
**Gauge:** 1435 mm (Gauge of 100.019-024 remains to be decided).
**Supply System:** 3000 V dc or 25 kV ac 50 Hz overhead.
**Normal Formation:** Motor + 8 Trailers + Motor
**Seats:** 108F, 212S. **Length:** 200.144 m.
**Electrical Equipment:** Alsthom. **Motors:** 8 of 1100 kW each.

**Wheel Arrangement:** Bo-Bo + 2-2-2-2-2-2-2-2-2 + Bo-Bo.
**Weight:** 393.00 tonnes.          **Maximum Speed:** 250 km/h.

| | | | | |
|---|---|---|---|---|
| 100.001 LS | 100.006 LS | 100.011 LS | 100.016 LS | 100.021 |
| 100.002 LS | 100.007 LS | 100.012 LS | 100.017 | 100.022 |
| 100.003 LS | 100.008 LS | 100.013 LS | 100.018 | 100.023 |
| 100.004 LS | 100.009 LS | 100.014 LS | 100.019 | 100.024 |
| 100.005 LS | 100.010 LS | 100.015 LS | 100.020 | |

# CLASS 432                                           3-Car Unit

**Built:** 1971-79 by CAF/Macosa/MMC/Cenemesa.
**Normal Formation:** BD + B + A.          **Seats:** 56F, 160S.
**Supply System:** 3000 v dc or 1500 V dc overhead.
**Electrical Equipment:** Melco.          **Length:** 79.870 m.
**Motors:** 4 of 290 kW each.          **Wheel Arrangement:** Bo-Bo + 2-2 + 2-2.
**Weight:** 150.40 tonnes.          **Maximum Speed:** 140 km/h.
**Former Numbers:** WMD 511-30 (432.001-020) in sequence.

| | | | | | |
|---|---|---|---|---|---|
| 432.001 | 432.004 | 432.009 | 432.012 | 432.015 | 432.018 |
| 432.002 | 432.006 | 432.010 | 432.013 | 432.016 | 432.019 |
| 432.003 | 432.008 | 432.011 | 432.014 | 432.017 | 432.020 |

# CLASS 439                                           2-Car Unit

**Built:** 1967-68 by CAF/Cenemesa/Metropolitan Cammell/ACEC.
**Normal Formation:** B + B.          **Seats:** 156S.
**Supply System:** 3000 V dc or 1500 V dc overhead.
**Electrical Equipment:**          **Length:** 51.910 m.
**Motors:** 4 of 252 kW each.          **Wheel Arrangement:** Bo-Bo + 2-2.
**Weight:** 96.80 tonnes.          **Maximum Speed:** 130 km/h.
**Former Numbers:** WMD 901-32 (439.001-032) in sequence.

| | | | |
|---|---|---|---|
| 439.003 | 439.004 | 439.005 | 439.006 |

# CLASS 440                                           3-Car Unit

**Built:** 1974-85 by CAF/Macosa/Westinghouse/GEE/Melco.
**Normal Formation:** BD + B + B.          **Seats:** 260S.
**Electrical Equipment:** Melco.          **Length:** 80.164 m.
**Motors:** 4 of 290 kW each.          **Wheel Arrangement:** Bo-Bo + 2-2 + 2-2.
**Weight:** 133.30 tonnes.          **Maximum Speed:** 140 km/h.

| | | | | | |
|---|---|---|---|---|---|
| 440.001 | 440.023 | 440.045 | 440.067 | 440.089 | 440.111 |
| 440.002 | 440.024 | 440.046 | 440.068 | 440.090 | 440.112 |
| 440.003 | 440.025 | 440.047 | 440.069 | 440.091 | 440.113 |
| 440.004 | 440.026 | 440.048 | 440.070 | 440.092 | 440.114 |
| 440.005 | 440.027 | 440.049 | 440.071 | 440.093 | 440.115 |
| 440.006 | 440.028 | 440.050 | 440.072 | 440.094 | 440.116 |
| 440.007 | 440.029 | 440.051 | 440.073 | 440.095 | 440.117 |
| 440.008 | 440.030 | 440.052 | 440.074 | 440.096 | 440.118 |
| 440.009 | 440.031 | 440.053 | 440.075 | 440.097 | 440.119 |
| 440.010 | 440.032 | 440.054 | 440.076 | 440.098 | 440.120 |
| 440.011 | 440.033 | 440.055 | 440.077 | 440.099 | 440.121 |
| 440.012 | 440.034 | 440.056 | 440.078 | 440.100 | 440.122 |
| 440.013 | 440.035 | 440.057 | 440.079 | 440.101 | 440.123 |
| 440.014 | 440.036 | 440.058 | 440.080 | 440.102 | 440.124 |
| 440.015 | 440.037 | 440.059 | 440.081 | 440.103 | 440.125 |
| 440.016 | 440.038 | 440.060 | 440.082 | 440.104 | 440.126 |
| 440.017 | 440.039 | 440.061 | 440.083 | 440.105 | 440.127 |
| 440.018 | 440.040 | 440.062 | 440.084 | 440.106 | 440.128 |
| 440.019 | 440.041 | 440.063 | 440.085 | 440.107 | 440.129 |
| 440.020 | 440.042 | 440.064 | 440.086 | 440.108 | 440.130 |
| 440.021 | 440.043 | 440.065 | 440.087 | 440.109 | 440.131 |
| 440.022 | 440.044 | 440.066 | 440.088 | 440.110 | 440.132 |

# SPAIN

| | | | | | |
|---|---|---|---|---|---|
| 440.133 | 440.154 | 440.175 | 440.196 | 440.216 | 440.236 |
| 440.134 | 440.155 | 440.176 | 440.197 | 440.217 | 440.237 |
| 440.135 | 440.156 | 440.177 | 440.198 | 440.218 | 440.238 |
| 440.136 | 440.157 | 440.178 | 440.199 | 440.219 | 440.239 |
| 440.137 | 440.158 | 440.179 | 440.200 | 440.220 | 440.240 |
| 440.138 | 440.159 | 440.180 | 440.201 | 440.221 | 440.241 |
| 440.139 | 440.160 | 440.181 | 440.202 | 440.222 | 440.242 |
| 440.140 | 440.161 | 440.182 | 440.203 | 440.223 | 440.243 |
| 440.141 | 440.162 | 440.183 | 440.204 | 440.224 | 440.244 |
| 440.142 | 440.163 | 440.184 | 440.205 | 440.225 | 440.245 |
| 440.143 | 440.164 | 440.185 | 440.206 | 440.226 | 440.246 |
| 440.144 | 440.165 | 440.186 | 440.207 | 440.227 | 440.247 |
| 440.145 | 440.166 | 440.187 | 440.208 | 440.228 | 440.248 |
| 440.146 | 440.167 | 440.188 | 440.209 | 440.229 | 440.249 |
| 440.147 | 440.168 | 440.189 | 440.210 | 440.230 | 440.250 |
| 440.148 | 440.169 | 440.190 | 440.211 | 440.231 | 440.251 |
| 440.149 | 440.170 | 440.191 | 440.212 | 440.232 | 440.252 |
| 440.150 | 440.171 | 440.192 | 440.213 | 440.233 | 440.253 |
| 440.151 | 440.172 | 440.193 | 440.214 | 440.234 | 440.501 |
| 440.152 | 440.173 | 440.194 | 440.215 | 440.235 | 440.502 |
| 440.153 | 440.174 | 440.195 | | | |

# CLASS 442      2-Car Unit

**Gauge:** 1000 mm.
**Built:** 1976-82 by MTM.
**Normal Formation:** B + B.
**Electrical Equipment:** BBC.
**Motors:** 4 of 131 kW each.
**Weight:** 63.20 tonnes.
**Note:** Work on the Cercedilla-Los Cotos route.

**Supply System:** 1500 V dc overhead.
**Seats:** 88S.
**Length:** 36.140 m.
**Wheel Arrangement:** Bo-Bo + 2-2.
**Maximum Speed:** 60 km/h.

| | | | | | |
|---|---|---|---|---|---|
| 442.001 | 442.002 | 442.003 | 442.004 | 442.005 | 442.006 |

# CLASS 444      3-Car Unit

**Built:** 1980-81 by CAF/Macosa/GEE/Westinghouse/Melco.
**Normal Formation:** BD + B + A.
**Electrical Equipment:** Melco.
**Motors:** 4 of 290 kW each.
**Weight:** 151.00 tonnes.

**Seats:** 52F, 160S.
**Length:** 79.864 m.
**Wheel Arrangement:** Bo-Bo + 2-2 + 2-2.
**Maximum Speed:** 140 km/h.

| | | | | | |
|---|---|---|---|---|---|
| 444.001 | 444.004 | 444.007 | 444.009 | 444.011 | 444.013 |
| 444.002 | 444.005 | 444.008 | 444.010 | 444.012 | 444.014 |
| 444.003 | 444.006 | | | | |

# CLASS 446      3-Car Unit

**Built:** 1989-93 by CAF/MTM/Meinfesa/Cenemesa.
**Normal Formation:** B + B + B.
**Electrical Equipment:** GEE.
**Motors:** 8 of 288 kW each.
**Weight:** 166.60 tonnes.

**Seats:** 242S.
**Length:** 75.993 m.
**Wheel Arrangement:** Bo-Bo + 2-2 + Bo-Bo.
**Maximum Speed:** 100 km/h.

| | | | | | |
|---|---|---|---|---|---|
| 9-446.001 | 7-446.001 | 9-446.002 | 9-446.023 | 7-446.012 | 9-446.024 |
| 9-446.003 | 7-446.002 | 9-446.004 | 9-446.025 | 7-446.013 | 9-446.026 |
| 9-446.005 | 7-446.003 | 9-446.006 | 9-446.027 | 7-446.014 | 9-446.028 |
| 9-446.007 | 7-446.004 | 9-446.008 | 9-446.029 | 7-446.015 | 9-446.030 |
| 9-446.009 | 7-446.005 | 9-446.010 | 9-446.031 | 7-446.016 | 9-446.032 |
| 9-446.011 | 7-446.006 | 9-446.012 | 9-446.033 | 7-446.017 | 9-446.034 |
| 9-446.013 | 7-446.007 | 9-446.014 | 9-446.035 | 7-446.018 | 9-446.036 |
| 9-446.015 | 7-446.008 | 9-446.016 | 9-446.037 | 7-446.019 | 9-446.038 |
| 9-446.017 | 7-446.009 | 9-446.018 | 9-446.039 | 7-446.020 | 9-446.040 |
| 9-446.019 | 7-446.010 | 9-446.020 | 9-446.041 | 7-446.021 | 9-446.042 |
| 9-446.021 | 7-446.011 | 9-446.022 | 9-446.043 | 7-446.022 | 9-446.044 |

| | | | | | |
|---|---|---|---|---|---|
| 9-446.045 | 7-446.023 | 9-446.046 | 9-446.171 | 7-446.086 | 9-446.172 |
| 9-446.047 | 7-446.024 | 9-446.048 | 9-446.173 | 7-446.087 | 9-446.174 |
| 9-446.049 | 7-446.025 | 9-446.050 | 9-446.175 | 7-446.088 | 9-446.176 |
| 9-446.051 | 7-446.026 | 9-446.052 | 9-446.177 | 7-446.089 | 9-446.178 |
| 9-446.053 | 7-446.027 | 9-446.054 | 9-446.179 | 7-446.090 | 9-446.180 |
| 9-446.055 | 7-446.028 | 9-446.056 | 9-446.181 | 7-446.091 | 9-446.182 |
| 9-446.057 | 7-446.029 | 9-446.058 | 9-446.183 | 7-446.092 | 9-446.184 |
| 9-446.059 | 7-446.030 | 9-446.060 | 9-446.185 | 7-446.093 | 9-446.186 |
| 9-446.061 | 7-446.031 | 9-446.062 | 9-446.187 | 7-446.094 | 9-446.188 |
| 9-446.063 | 7-446.032 | 9-446.064 | 9-446.189 | 7-446.095 | 9-446.190 |
| 9-446.065 | 7-446.033 | 9-446.066 | 9-446.191 | 7-446.096 | 9-446.192 |
| 9-446.067 | 7-446.034 | 9-446.068 | 9-446.193 | 7-446.097 | 9-446.194 |
| 9-446.069 | 7-446.035 | 9-446.070 | 9-446.195 | 7-446.098 | 9-446.196 |
| 9-446.071 | 7-446.036 | 9-446.072 | 9-446.197 | 7-446.099 | 9-446.198 |
| 9-446.073 | 7-446.037 | 9-446.074 | 9-446.199 | 7-446.100 | 9-446.200 |
| 9-446.075 | 7-446.038 | 9-446.076 | 9-446.201 | 7-446.101 | 9-446.202 |
| 9-446.077 | 7-446.039 | 9-446.078 | 9-446.203 | 7-446.102 | 9-446.204 |
| 9-446.079 | 7-446.040 | 9-446.080 | 9-446.205 | 7-446.103 | 9-446.206 |
| 9-446.081 | 7-446.041 | 9-446.082 | 9-446.207 | 7-446.104 | 9-446.208 |
| 9-446.083 | 7-446.042 | 9-446.084 | 9-446.209 | 7-446.105 | 9-446.210 |
| 9-446.085 | 7-446.043 | 9-446.086 | 9-446.211 | 7-446.106 | 9-446.212 |
| 9-446.087 | 7-446.044 | 9-446.088 | 9-446.213 | 7-446.107 | 9-446.214 |
| 9-446.089 | 7-446.045 | 9-446.090 | 9-446.215 | 7-446.108 | 9-446.216 |
| 9-446.091 | 7-446.046 | 9-446.092 | 9-446.217 | 7-446.109 | 9-446.218 |
| 9-446.093 | 7-446.047 | 9-446.094 | 9-446.219 | 7-446.110 | 9-446.220 |
| 9-446.095 | 7-446.048 | 9-446.096 | 9-446.221 | 7-446.111 | 9-446.222 |
| 9-446.097 | 7-446.049 | 9-446.098 | 9-446.223 | 7-446.112 | 9-446.224 |
| 9-446.099 | 7-446.050 | 9-446.100 | 9-446.225 | 7-446.113 | 9-446.226 |
| 9-446.101 | 7-446.051 | 9-446.102 | 9-446.227 | 7-446.114 | 9-446.228 |
| 9-446.103 | 7-446.052 | 9-446.104 | 9-446.229 | 7-446.115 | 9-446.230 |
| 9-446.105 | 7-446.053 | 9-446.106 | 9-446.231 | 7-446.116 | 9-446.232 |
| 9-446.107 | 7-446.054 | 9-446.108 | 9-446.233 | 7-446.117 | 9-446.234 |
| 9-446.109 | 7-446.055 | 9-446.110 | 9-446.235 | 7-446.118 | 9-446.236 |
| 9-446.111 | 7-446.056 | 9-446.112 | 9-446.237 | 7-446.119 | 9-446.238 |
| 9-446.113 | 7-446.057 | 9-446.114 | 9-446.239 | 7-446.120 | 9-446.240 |
| 9-446.115 | 7-446.058 | 9-446.116 | 9-446.241 | 7-446.121 | 9-446.242 |
| 9-446.117 | 7-446.059 | 9-446.118 | 9-446.243 | 7-446.122 | 9-446.244 |
| 9-446.119 | 7-446.060 | 9-446.120 | 9-446.245 | 7-446.123 | 9-446.246 |
| 9-446.121 | 7-446.061 | 9-446.122 | 9-446.247 | 7-446.124 | 9-446.248 |
| 9-446.123 | 7-446.062 | 9-446.124 | 9-446.249 | 7-446.125 | 9-446.250 |
| 9-446.125 | 7-446.063 | 9-446.126 | 9-446.251 | 7-446.126 | 9-446.252 |
| 9-446.127 | 7-446.064 | 9-446.128 | 9-446.253 | 7-446.127 | 9-446.254 |
| 9-446.129 | 7-446.065 | 9-446.130 | 9-446.255 | 7-446.128 | 9-446.256 |
| 9-446.131 | 7-446.066 | 9-446.132 | 9-446.257 | 7-446.129 | 9-446.258 |
| 9-446.133 | 7-446.067 | 9-446.134 | 9-446.259 | 7-446.130 | 9-446.260 |
| 9-446.135 | 7-446.068 | 9-446.136 | 9-446.261 | 7-446.131 | 9-446.262 |
| 9-446.137 | 7-446.069 | 9-446.138 | 9-446.263 | 7-446.132 | 9-446.264 |
| 9-446.139 | 7-446.070 | 9-446.140 | 9-446.265 | 7-446.133 | 9-446.266 |
| 9-446.141 | 7-446.071 | 9-446.142 | 9-446.267 | 7-446.134 | 9-446.268 |
| 9-446.143 | 7-446.072 | 9-446.144 | 9-446.269 | 7-446.135 | 9-446.270 |
| 9-446.145 | 7-446.073 | 9-446.146 | 9-446.271 | 7-446.136 | 9-446.272 |
| 9-446.147 | 7-446.074 | 9-446.148 | 9-446.273 | 7-446.137 | 9-446.274 |
| 9-446.149 | 7-446.075 | 9-446.150 | 9-446.275 | 7-446.138 | 9-446.276 |
| 9-446.151 | 7-446.076 | 9-446.152 | 9-446.277 | 7-446.139 | 9-446.278 |
| 9-446.153 | 7-446.077 | 9-446.154 | 9-446.279 | 7-446.140 | 9-446.280 |
| 9-446.155 | 7-446.078 | 9-446.156 | 9-446.281 | 7-446.141 | 9-446.282 |
| 9-446.157 | 7-446.079 | 9-446.158 | 9-446.283 | 7-446.142 | 9-446.284 |
| 9-446.159 | 7-446.080 | 9-446.160 | 9-446.285 | 7-446.143 | 9-446.286 |
| 9-446.161 | 7-446.081 | 9-446.162 | 9-446.287 | 7-446.144 | 9-446.288 |
| 9-446.163 | 7-446.082 | 9-446.164 | 9-446.289 | 7-446.145 | 9-446.290 |
| 9-446.165 | 7-446.083 | 9-446.166 | 9-446.291 | 7-446.146 | 9-446.292 |
| 9-446.167 | 7-446.084 | 9-446.168 | 9-446.293 | 7-446.147 | 9-446.294 |
| 9-446.169 | 7-446.085 | 9-446.170 | 9-446.295 | 7-446.148 | 9-446.296 |

# SPAIN

| | | | | | |
|---|---|---|---|---|---|
| 9-446.297 | 7-446.149 | 9-446.298 | 9-446.319 | 7-446.160 | 9-446.320 |
| 9-446.299 | 7-446.150 | 9-446.300 | 9-446.321 | 7-446.161 | 9-446.322 |
| 9-446.301 | 7-446.151 | 9-446.302 | 9-446.323 | 7-446.162 | 9-446.324 |
| 9-446.303 | 7-446.152 | 9-446.304 | 9-446.325 | 7-446.163 | 9-446.326 |
| 9-446.305 | 7-446.153 | 9-446.306 | 9-446.327 | 7-446.164 | 9-446.328 |
| 9-446.307 | 7-446.154 | 9-446.308 | 9-446.329 | 7-446.165 | 9-446.330 |
| 9-446.309 | 7-446.155 | 9-446.310 | 9-446.331 | 7-446.166 | 9-446.332 |
| 9-446.311 | 7-446.156 | 9-446.312 | 9-446.333 | 7-446.167 | 9-446.334 |
| 9-446.313 | 7-446.157 | 9-446.314 | 9-446.335 | 7-446.168 | 9-446.336 |
| 9-446.315 | 7-446.158 | 9-446.316 | 9-446.337 | 7-446.169 | 9-446.338 |
| 9-446.317 | 7-446.159 | 9-446.318 | 9-446.339 | 7-446.170 | 9-446.340 |

# CLASS 447                                                    3-Car Unit

**Built:** 1992-94 by CAF/Siemens/Cenemesa.
**Normal Formation:** B + B + B.                **Seats:** 242S.
**Electrical Equipment:** Siemens.              **Length:** 75.993 m.
**Motors:** 8 of 300 kW each.                   **Wheel Arrangement:** Bo-Bo + 2-2 + Bo-Bo.
**Weight:** 157.00 tonnes.                      **Maximum Speed:** 120 km/h.

| | | | | | |
|---|---|---|---|---|---|
| 9-447.001 | 7-447.001 | 9-447.002 | 9-447.073 | 7-447.037 | 9-447.074 |
| 9-447.003 | 7-447.002 | 9-447.004 | 9-447.075 | 7-447.038 | 9-447.076 |
| 9-447.005 | 7-447.003 | 9-447.006 | 9-447.077 | 7-447.039 | 9-447.078 |
| 9-447.007 | 7-447.004 | 9-447.008 | 9-447.079 | 7-447.040 | 9-447.080 |
| 9-447.009 | 7-447.005 | 9-447.010 | 9-447.081 | 7-447.041 | 9-447.082 |
| 9-447.011 | 7-447.006 | 9-447.012 | 9-447.083 | 7-447.042 | 9-447.084 |
| 9-447.013 | 7-447.007 | 9-447.014 | 9-447.085 | 7-447.043 | 9-447.086 |
| 9-447.015 | 7-447.008 | 9-447.016 | 9-447.087 | 7-447.044 | 9-447.088 |
| 9-447.017 | 7-447.009 | 9-447.018 | 9-447.089 | 7-447.045 | 9-447.090 |
| 9-447.019 | 7-447.010 | 9-447.020 | 9-447.091 | 7-447.046 | 9-447.092 |
| 9-447.021 | 7-447.011 | 9-447.022 | 9-447.093 | 7-447.047 | 9-447.094 |
| 9-447.023 | 7-447.012 | 9-447.024 | 9-447.095 | 7-447.048 | 9-447.096 |
| 9-447.025 | 7-447.013 | 9-447.026 | 9-447.097 | 7-447.049 | 9-447.098 |
| 9-447.027 | 7-447.014 | 9-447.028 | 9-447.099 | 7-447.050 | 9-447.100 |
| 9-447.029 | 7-447.015 | 9-447.030 | 9-447.101 | 7-447.051 | 9-447.102 |
| 9-447.031 | 7-447.016 | 9-447.032 | 9-447.103 | 7-447.052 | 9-447.104 |
| 9-447.033 | 7-447.017 | 9-447.034 | 9-447.105 | 7-447.053 | 9-447.106 |
| 9-447.035 | 7-447.018 | 9-447.036 | 9-447.107 | 7-447.054 | 9-447.108 |
| 9-447.037 | 7-447.019 | 9-447.038 | 9-447.109 | 7-447.055 | 9-447.110 |
| 9-447.039 | 7-447.020 | 9-447.040 | 9-447.111 | 7-447.056 | 9-447.112 |
| 9-447.041 | 7-447.021 | 9-447.042 | 9-447.113 | 7-447.057 | 9-447.114 |
| 9-447.043 | 7-447.022 | 9-447.044 | 9-447.115 | 7-447.058 | 9-447.116 |
| 9-447.045 | 7-447.023 | 9-447.046 | 9-447.117 | 7-447.059 | 9-447.118 |
| 9-447.047 | 7-447.024 | 9-447.048 | 9-447.119 | 7-447.060 | 9-447.120 |
| 9-447.049 | 7-447.025 | 9-447.050 | 9-447.121 | 7-447.061 | 9-447.122 |
| 9-447.051 | 7-447.026 | 9-447.052 | 9-447.123 | 7-447.062 | 9-447.124 |
| 9-447.053 | 7-447.027 | 9-447.054 | 9-447.125 | 7-447.063 | 9-447.126 |
| 9-447.055 | 7-447.028 | 9-447.056 | 9-447.127 | 7-447.064 | 9-447.128 |
| 9-447.057 | 7-447.029 | 9-447.058 | 9-447.129 | 7-447.065 | 9-447.130 |
| 9-447.059 | 7-447.030 | 9-447.060 | 9-447.131 | 7-447.066 | 9-447.132 |
| 9-447.061 | 7-447.031 | 9-447.062 | 9-447.133 | 7-447.067 | 9-447.134 |
| 9-447.063 | 7-447.032 | 9-447.064 | 9-447.135 | 7-447.068 | 9-447.136 |
| 9-447.065 | 7-447.033 | 9-447.066 | 9-447.137 | 7-447.069 | 9-447.138 |
| 9-447.067 | 7-447.034 | 9-447.068 | 9-447.139 | 7-447.070 | 9-447.140 |
| 9-447.069 | 7-447.035 | 9-447.070 | 9-447.141 | 7-447.071 | 9-447.142 |
| 9-447.071 | 7-447.036 | 9-447.072 | | | |

# CLASS 448                                                    3-Car Unit

**Built:** 1987-1991 by CAF/Macosa/Cenemesa/Atiensa/MTM/Westinghouse.
**Normal Formation:** BD + B + A.               **Seats:** 72F, 140S.
**Electrical Equipment:** Melco.                **Length:** 80.044 m.
**Motors:** 4 of 290 kW each.                   **Wheel Arrangement:** Bo-Bo+2-2 +2-2.
**Weight:** 151.00 tonnes.                      **Maximum Speed:** 160 km/h.

| | | | | | |
|---|---|---|---|---|---|
| 448.001 | 448.007 | 448.012 | 448.017 | 448.022 | 448.027 |
| 448.002 | 448.008 | 448.013 | 448.018 | 448.023 | 448.028 |
| 448.003 | 448.009 | 448.014 | 448.019 | 448.024 | 448.029 |
| 448.004 | 448.010 | 448.015 | 448.020 | 448.025 | 448.030 |
| 448.005 | 448.011 | 448.016 | 448.021 | 448.026 | 448.031 |
| 448.006 | | | | | |

## CLASS 450                                        6-Car Unit

**Built:** 1993-95 by CAF/GEC Alsthom.
**Normal Formation:** B + B + B + B + B + B.
**Electrical Equipment:** GEC Alsthom.
**Motors:** 8 of 370 kW each.
**Weight:** 350.40 tonnes.

**Seats:** 1008S.
**Length:** 159.400 m.
**Wheel Arrangement:** Bo-Bo + 2-2 + 2-2 + 2-2 + 2-2 + Bo-Bo.
**Maximum Speed:** 140 km/h.

| | | | | | |
|---|---|---|---|---|---|
| 9-450.001 | 7-450.201 | 7-450.202 | 7-450.001 | 7-450.203 | 9-450.002 |
| 9-450.003 | 7-450.204 | 7-450.205 | 7-450.002 | 7-450.206 | 9-450.004 |
| 9-450.005 | 7-450.207 | 7-450.208 | 7-450.003 | 7-450.209 | 9-450.006 |
| 9-450.007 | 7-450.210 | 7-450.211 | 7-450.004 | 7-450.212 | 9-450.008 |
| 9-450.009 | 7-450.213 | 7-450.214 | 7-450.005 | 7-450.215 | 9-450.010 |
| 9-450.011 | 7-450.216 | 7-450.217 | 7-450.006 | 7-450.218 | 9-450.012 |
| 9-450.013 | 7-450.219 | 7-450.220 | 7-450.007 | 7-450.221 | 9-450.014 |
| 9-450.015 | 7-450.222 | 7-450.223 | 7-450.008 | 7-450.224 | 9-450.016 |
| 9-450.017 | 7-450.225 | 7-450.226 | 7-450.009 | 7-450.227 | 9-450.018 |
| 9-450.019 | 7-450.228 | 7-450.229 | 7-450.010 | 7-450.230 | 9-450.020 |
| 9-450.021 | 7-450.231 | 7-450.232 | 7-450.011 | 7-450.233 | 9-450.022 |
| 9-450.023 | 7-450.234 | 7-450.235 | 7-450.012 | 7-450.236 | 9-450.024 |
| 9-450.025 | 7-450.237 | 7-450.238 | 7-450.013 | 7-450.239 | 9-450.026 |
| 9-450.027 | 7-450.240 | 7-450.241 | 7-450.014 | 7-450.242 | 9-450.028 |
| 9-450.029 | 7-450.243 | 7-450.244 | 7-450.015 | 7-450.245 | 9-450.030 |
| 9-450.031 | 7-450.246 | 7-450.247 | 7-450.016 | 7-450.248 | 9-450.032 |
| 9-450.033 | 7-450.249 | 7-450.250 | 7-450.017 | 7-450.251 | 9-450.034 |
| 9-450.035 | 7-450.252 | 7-450.253 | 7-450.018 | 7-450.254 | 9-450.036 |
| 9-450.037 | 7-450.255 | 7-450.256 | 7-450.019 | 7-450.257 | 9-450.038 |
| 9-450.039 | 7-450.258 | 7-450.259 | 7-450.020 | 7-450.260 | 9-450.040 |
| 9-450.041 | 7-450.261 | 7-450.262 | 7-450.021 | 7-450.263 | 9-450.042 |
| 9-450.043 | 7-450.264 | 7-450.265 | 7-450.022 | 7-450.266 | 9-450.044 |
| 9-450.045 | 7-450.267 | 7-450.268 | 7-450.023 | 7-450.269 | 9-450.046 |
| 9-450.047 | 7-450.270 | 7-450.271 | 7-450.024 | 7-450.272 | 9-450.048 |

## CLASS 451                                        3-Car Unit

**Built:** 1994-95 by CAF/GEC Alsthom.
**Normal Formation:** B + B + B .
**Electrical Equipment:** GEC Alsthom.
**Motors:** 4 of 370 kW each.
**Weight:** 175.20 tonnes.

**Seats:** 504S.
**Length:** 79.700 m.
**Wheel Arrangement:** Bo-Bo + 2-2 + 2-2.
**Maximum Speed:** 140 km/h.

| | | | | | |
|---|---|---|---|---|---|
| 451.001 | 451.003 | 451.005 | 451.007 | 451.009 | 451.011 |
| 451.002 | 451.004 | 451.006 | 451.008 | 451.010 | 451.012 |

# DIESEL MULTIPLE UNITS

## CLASS 592                                        3-Car Unit

**Built:** 1981-84 by Macosa/Atiensa/MAN.
**Normal Formation:** B + B + B.                    **Seats:** 228S.
**Engines:** Two MAN D3256 of 115 kW at 2100 rpm per power car.
**Transmission:** Hydraulic. Voith L211r.
**Length:** 70.214 m.                               **Wheel Arrangement:** 1A-A1 + 2-2 + 1A-A1.
**Weight:** 131.00 tonnes.                          **Maximum Speed:** 120 km/h.
**Note:** 9-592 301 is a single engined motor coach converted from a trailer. Maximum speed of this unit is 140 km/h.

# SPAIN

◀ FEVE 1400 Class 1401 refuelling at Oviedo (E. Barnes)

▶ FEVE 2400 Class 2-car unit 2467 + 2417 at Nava during May 1993 (B. Philpott)

◀ RENFE AVE set 100.001 at Córdoba with the 1400 Sevilla - Madrid service (B. Philpott)

▶ Refurbished RENFE Class 440 unit 440.176 at Posadas on a Córdoba - Sevilla service on 16.06.94 (B. Philpott)

| | | | | | | |
|---|---|---|---|---|---|---|
| 9-592.001 | 7-592.001 | 9-592.002 | | 9-592.039 | 7-592.020 | 9-592.040 |
| 9-592.003 | 7-592.002 | 9-592.004 | | 9-592.041 | 7-592.021 | 9-592.042 |
| 9-592.005 | 7-592.003 | 9-592.006 | | 9-592.043 | 7-592.022 | 9-592.044 |
| 9-592.007 | 7-592.004 | 9-592.008 | | 9-592.045 | 7-592.023 | 9-592.046 |
| 9-592.009 | 7-592.005 | 9-592.010 | | 9-592.047 | 7-592.024 | 9-592.048 |
| 9-592.011 | 7-592.006 | 9-592.012 | | 9-592.049 | 7-592.025 | 9-592.050 |
| 9-592.013 | 7-592.007 | 9-592.014 | | 9-592.051 | 7-592.026 | 9-592.052 |
| 9-592.015 | 7-592.008 | 9-592.016 | | 9-592.053 | 7-592.027 | 9-592.054 |
| 9-592.017 | 7-592.009 | 9-592.018 | | 9-592.055 | 7-592.028 | 9-592.056 |
| 9-592.021 | 7-592.011 | 9-592.022 | | 9-592.057 | 7-592.029 | 9-592.058 |
| 9-592.023 | 7-592.012 | 9-592.024 | | 9-592.059 | 7-592.030 | 9-592.060 |
| 9-592.025 | 7-592.013 | 9-592.026 | | 9-592.061 | 7-592.031 | 9-592.062 |
| 9-592.027 | 7-592.014 | 9-592.028 | | 9-592.063 | 7-592.032 | 9-592.064 |
| 9-592.029 | 7-592.015 | 9-592.030 | | 9-592.065 | 7-592.033 | 9-592.066 |
| 9-592.031 | 7-592.016 | 9-592.032 | | 9-592.067 | 7-592.034 | 9-592.068 |
| 9-592.033 | 7-592.017 | 9-592.034 | | 9-592.069 | 7-592.035 | 9-592.070 |
| 9-592.035 | 7-592.018 | 9-592.036 | | 9-592.301 | | 9-592.302 |
| 9-592.037 | 7-592.019 | 9-592.038 | | | | |

## CLASS 593 — 3-Car Unit

**Built:** 1982-84 by CAF/Babcock/Fiat.
**Normal Formation:** B + B + B.          **Seats:** 228S.
**Engines:** Two Fiat 8217.32 of 123 kW at 2000 rpm per power car.
**Transmission:** Mechanical.
**Length:** 71.420 m.          **Wheel Arrangement:** 1A-A1 + 2-2 + 1A-A1.
**Weight:** 136.00 tonnes.          **Maximum Speed:** 120 km/h.

| | | | | | | |
|---|---|---|---|---|---|---|
| 9-593.001 | 7-593.001 | 9-593.002 | | 9-593.033 | 7-593.017 | 9-593.034 |
| 9-593.003 | 7-593.002 | 9-593.004 | | 9-593.035 | 7-593.018 | 9-593.036 |
| 9-593.005 | 7-593.003 | 9-593.006 | | 9-593.037 | 7-593.019 | 9-593.038 |
| 9-593.007 | 7-593.004 | 9-593.008 | | 9-593.039 | 7-593.020 | 9-593.040 |
| 9-593.009 | 7-593.005 | 9-593.010 | | 9-593.041 | 7-593.021 | 9-593.042 |
| 9-593.011 | 7-593.006 | 9-593.012 | | 9-593.043 | 7-593.022 | 9-593.044 |
| 9-593.013 | 7-593.007 | 9-593.014 | | 9-593.045 | 7-593.023 | 9-593.046 |
| 9-593.015 | 7-593.008 | 9-593.016 | | 9-593.047 | 7-593.024 | 9-593.048 |
| 9-593.017 | 7-593.009 | 9-593.018 | | 9-593.049 | 7-593.025 | 9-593.050 |
| 9-593.019 | 7-593.010 | 9-593.020 | | 9-593.051 | 7-593.026 | 9-593.052 |
| 9-593.021 | 7-593.011 | 9-593.022 | | 9-593.053 | 7-593.027 | 9-593.054 |
| 9-593.023 | 7-593.012 | 9-593.024 | | 9-593.055 | 7-593.028 | 9-593.056 |
| 9-593.025 | 7-593.013 | 9-593.026 | | 9-593.057 | 7-593.029 | 9-593.058 |
| 9-593.027 | 7-593.014 | 9-593.028 | | 9-593.059 | 7-593.030 | 9-593.060 |
| 9-593.029 | 7-593.015 | 9-593.030 | | 9-593.061 | 7-593.031 | 9-593.062 |
| 9-593.031 | 7-593.016 | 9-593.032 | | | | |

## CLASS 597 — 2-Car Unit

**Built:** 1964-66 by CAF/MMC/Fiat.
**Normal Formation:** B + B.          **Seats:** 128S.
**Engine:** OM Saurer SEHL of 381 kW.          **Transmission:** Hydraulic. Fiat.
**Length:** 53.200 m.          **Wheel Arrangement:** 2-B + 2-2.
**Weight:** 96.80 tonnes.          **Maximum Speed:** 120 km/h.

| | | | | | | |
|---|---|---|---|---|---|---|
| 597.003 | 597.010 | 597.017 | | 597.029 | 597.043 | 597.048 |
| 597.004 | 597.011 | 597.018 | | 597.036 | 597.044 | 597.052 |
| 597.009 | | | | | | |

# Ferrocarriles Españoles de Via Estrecha (FEVE).

**Gauge:** 1000 mm.          **Route Length:** 1221 km.
**Electrification System:** 1500 V dc overhead.
**Depots:**

# DIESEL LOCOMOTIVES

## CLASS 1000             Bo-Bo

**Built:** 1955-65 by Alsthom/Euskalduna/Babcock.
**Engine:** SACM MGO V12 SHR of 590 kW at 1500 rpm.
**Transmission:** Electric.
**Weight:** 44.00 tonnes.          **Length:** 11.174 m.
**Maximum Speed:** 70 km/h.       **Train Supply:** Not equipped.
**Note:** Various withdrawn locomotives of this class are still extant.

| 1021 | 1023 | 1031 |
|------|------|------|

## CLASS 1050             Bo-Bo

**Built:** 1955-65 by Alsthom/Euskalduna/Babcock.
**Engine:** SACM MGO V12 ASHR of 625 kW at 1500 rpm.
**Transmission:** Electric.
**Weight:** 44.00 tonnes.          **Length:** 11.174 m.
**Maximum Speed:** 70 km/h.       **Train Supply:** Not equipped.
**Note:** Various withdrawn locomotives of this class are still extant.

| 1051 | 1059 |
|------|------|

## CLASS 1400             B-B

**Built:** 1964-66 by Henschel.
**Engine:** SACM BZSHR of 900 kW.     **Transmission:** Hydraulic.
**Weight:** 56.00 tonnes.          **Length:** 12.700 m.
**Maximum Speed:** 60 km/h.       **Train Supply:** Not equipped.

| 1401 | 1402 | 1403 | 1405 |
|------|------|------|------|

## CLASS 1500             Bo-Bo

**Built:** 1965 by General Electric.
**Engine:** Catterpillar 398B of 785 kW at 1300 rpm.
**Transmission:** Electric.
**Weight:** 52.595 tonnes.         **Length:** 11.329 m.
**Maximum Speed:** 80 km/h.       **Train Supply:** Not equipped.

| 1501 | 1503 | 1505 | 1507 | 1509 | 1510 |
|------|------|------|------|------|------|
| 1502 | 1504 | 1506 | 1508 |      |      |

## CLASS 1600             Bo-Bo

**Built:** 1982-83 by MTM/Alsthom.
**Engine:** SACM MGO V16 BZSHR of 1180 kW at 1500 rpm.
**Transmission:** Electric.
**Weight:** 54.00 tonnes.          **Length:** 13.324 m.
**Maximum Speed:** 70 km/h.       **Train Supply:** Not equipped.

| 1601 | 1604 | 1607 | 1609 | 1611 | 1613 |
|------|------|------|------|------|------|
| 1602 | 1605 | 1608 | 1610 | 1612 | 1614 |
| 1603 | 1606 |      |      |      |      |

## CLASS 1650             Bo-Bo

**Built:** 1985 by MTM/Alsthom.
**Engine:** SACM MGO V16 BZSHR of 1180 kW at 1500 rpm.
**Transmission:** Electric.
**Weight:** 54.00 tonnes.          **Length:** 14.000 m.
**Maximum Speed:** 70 km/h.       **Train Supply:** Not equipped.

| 1651 | 1653 | 1655 | 1657 | 1659 | 1660 |
|------|------|------|------|------|------|
| 1652 | 1654 | 1656 | 1658 |      |      |

# DIESEL MULTIPLE UNITS

## CLASS 2300 — Single Unit

**Built:** 1966-73 by MAN/CAF/Euskalduna/Ateinsa. Rebuilt 1984 by CAF/Babcock/Macosa.
**Normal Formation:** B.  **Seats:** 44S.
**Engines:** One Pegaso 9152/2 of 155 kW.  **Transmission:** Hydraulic. Voith Diwabus.
**Length:** 17.500 m.  **Wheel Arrangement:** B-2.
**Weight:** 23.20 tonnes.  **Maximum Speed:** 80 km/h.
**Note:** Used on the Cartagena-Los Nietos Pescaderia route.

| | | | | |
|---|---|---|---|---|
| 2309 | 2326 | 2328 | 2367 | 2372 |

## CLASS 2300 — 2-Car Unit

**Built:** 1966-73 by MAN/CAF/Euskalduna/Ateinsa. Rebuilt 1984 by CAF/Babcock/Macosa.
**Normal Formation:** B + B.  **Seats:** 88S.
**Engines:** Two Volvo THD 101 GB of 163 kW each.  **Transmission:** Hydraulic. Voith Diwabus.
**Length:** 34.920 m.  **Wheel Arrangement:** B-2 + 2-B.
**Weight:** 46.00 tonnes.  **Maximum Speed:** 80 km/h.
**Note:** Used on the Bilbao-Valmaseda (* Oviedo-Fuso de la Reina-Pravia/Collanzo) route.

| | | | | | | | |
|---|---|---|---|---|---|---|---|
| 2301 | 2347* | 2310 | 2318 | 2324 | 2366* | 2337 | 2361 |
| 2302 | 2331* | 2311 | 2325 | 2327 | 2338 | 2340 | 2369* |
| 2303 | 2315 | 2319 | 2356* | 2330 | 2343 | 2345 | 2355* |
| 2305 | 2322* | 2321 | 2371 | 2333 | 2336 | 2350 | 2353* |
| 2307 | 2351* | 2323 | 2352* | 2335 | 2339* | 2370 | 2373* |
| 2308 | 2341* | | | | | | |

## CLASS 2400 — 3-Car Unit

**Built:** 1983-84 by MTM.
**Normal Formation:** B + B + B.  **Seats:** 101S.
**Engines:** Two MAN D-3256 BTYVE of 228 kW each at 2100 rpm.
**Transmission:** Electric. BBC.
**Length:** 51.246 m.  **Wheel Arrangement:** Bo-2 + 2-2 + 2-Bo.
**Weight:** 89.85 tonnes.  **Maximum Speed:** 100 km/h.
**Note:** Used on non electrified lines in the north of Spain.

| | | | | | | | | |
|---|---|---|---|---|---|---|---|---|
| 2401 | 5401 | 2451 | 2405 | 5405 | 2455 | 2409 | 5409 | 2459 |
| 2402 | 5402 | 2452 | 2406 | 5406 | 2456 | 2410 | 5410 | 2460 |
| 2403 | 5403 | 2453 | 2407 | 5407 | 2457 | 2411 | 5411 | 2461 |
| 2404 | 5404 | 2454 | 2408 | 5408 | 2458 | 2413 | 5413 | 2463 |

## CLASS 2400 — 2-Car Unit

**Built:** 1985-86 by MTM/Macosa.
**Normal Formation:** B + B.  **Seats:** 55S.
**Engines:** Two MAN D-3256 BTYVE of 228 kW each at 2100 rpm.
**Transmission:** Electric. BBC.
**Length:** 34.300 m.  **Wheel Arrangement:** Bo-2 + 2-Bo.
**Weight:** 67.75 tonnes.  **Maximum Speed:** 100 km/h.
**Note:** Used on non electrified lines in the north of Spain.

| | | | | | | | |
|---|---|---|---|---|---|---|---|
| 2415 | 2465 | 2419 | 2469 | 2423 | 2473 | 2427 | 2477 |
| 2416 | 2466 | 2420 | 2470 | 2424 | 2474 | 2428 | 2478 |
| 2417 | 2467 | 2421 | 2471 | 2425 | 2475 | 2429 | 2479 |
| 2418 | 2468 | 2422 | 2472 | 2426 | 2476 | | |

## CLASS 2500 — 2-Car Unit

**Built:** 1988 by MTM.
**Normal Formation:** B + B.  **Seats:** 80S.

# SPAIN

**Engines:** Two Pegaso 96 RAUZ of 250 kW each at 2100 rpm.
**Transmission:** Three Phase Electric.
**Length:** 34.154 m.
**Weight:** 35.60 tonnes.

**Wheel Arrangement:** Bo-2 + 2-Bo.
**Maximum Speed:** 80 km/h.

2501 2551

# ELECTRIC MULTIPLE UNITS

## CLASS 3500          3-Car Unit

**Built:** 1977-84 by CAF.
**Normal Formation:** B + B + B.
**Electrical Equipment:** AEG/GEE.
**Motors:** 4 of 122 kW each.
**Weight:** 72.50 tonnes.

**Seats:** 100S.
**Length:** 47.710 m.
**Wheel Arrangement:** Bo-Bo + 2-2 + 2-2.
**Maximum Speed:** 80 km/h.
**Note:** Used on the Santander-Torrelavega-Cabezon de la Sal & Santander-Lierganes routes.

| | | | | | | | | |
|---|---|---|---|---|---|---|---|---|
| 3516 | 6551 | 6516 | 3524 | 6559 | 6524 | 3531 | 6566 | 6531 |
| 3517 | | 6517 | 3525 | 6560 | 6525 | 3532 | 6567 | 6532 |
| 3518 | 6553 | 6518 | 3526 | 6561 | 6526 | 3533 | 6568 | 6533 |
| 3519 | 6554 | 6519 | 3527 | 6562 | 6527 | 3534 | 6569 | 6534 |
| 3520 | | 6520 | 3528 | 6563 | 6528 | 3535 | 6570 | 6535 |
| 3521 | 6556 | 6521 | 3529 | 6564 | 6529 | 3536 | 6571 | 6536 |
| 3522 | 6557 | 6522 | 3530 | 6565 | 6530 | 3537 | 6572 | 6537 |
| 3523 | 6558 | 6523 | | | | | | |

## CLASS 3800          2-Car Unit

**Built:** 1991-93 by CAF.
**Normal Formation:** B + B.
**Electrical Equipment:** AEG/ABB.
**Motors:** 4 of 129 kW each.
**Weight:** 57.00 tonnes.

**Seats:** 87S.
**Length:** 32.000 m.
**Wheel Arrangement:** Bo-Bo + 2-2.
**Maximum Speed:** 80 km/h.
**Note:** Used on the electrified routes from Gijon.

| | | | | | | | |
|---|---|---|---|---|---|---|---|
| 3801 | 6801 | 3805 | 6805 | 3809 | 6809 | 3813 | 6813 |
| 3802 | 6802 | 3806 | 6806 | 3810 | 6810 | 3814 | 6814 |
| 3803 | 6803 | 3807 | 6807 | 3811 | 6811 | 3815 | 6815 |
| 3804 | 6804 | 3808 | 6808 | 3812 | 6812 | 3816 | 6816 |

# Eusko Trenbideak/Ferrocarriles Vascos (ET/FV)

**Gauge:** 1000 mm.
**Electrification System:** 1500 V dc overhead.
**Depots:**

**Route Length:** 202 km.

# ELECTRIC LOCOMOTIVES

## 11-14          Bo-Bo

**Built:** 1931-32 by CAF/ASEA.
**Electrical Equipment:** ASEA.
**Weight:** 43.00 tonnes.
**Maximum Speed:** 70 km/h.

**Continuous Rating:** 530 kW.
**Length:** 10.580 m.

11         12         13         14

## 17          Bo-Bo

**Built:** 1950 by ASEA. Rebuilt 1994 onwards by ET/FV Durango Works.
**Electrical Equipment:** ASEA.
**Continuous Rating:** 530 kW.

**Weight:** 44.00 tonnes.  **Length:** 10.850 m.
**Maximum Speed:** 70 km/h.

| 15 | *Amboto* | 16 | Oiz | 17 | *Udalaitz* |

# ELECTRIC MULTIPLE UNITS

## CLASS MAB                                          Single Unit

**Built:** 1928. Rebuilt 1950 & 1990 by ET/FV Sopelana Works.
**Normal Formation:** B.                    **Seats:** 32.
**Electrical Equipment:** AEG.              **Length:** 16.250 m.
**Motors:** 4 of 52 kW each.                **Wheel Arrangement:** Bo-Bo.
**Weight:** 32.80 tonnes.                   **Maximum Speed:** 60 km/h.
**Note:** For use on the Lutxana-Asua-Sondika line.

MAB 2        MAB 13

## CLASS 200                                          3-Car Unit

**Built:** 1985-86 by CAF/Babcock/GEE.
**Normal Formation:** B + B + B.           **Seats:** 192.
**Electrical Equipment:** GEE.             **Length:** 51.650 m.
**Motors:** 8 of 163 kW each.              **Wheel Arrangement:** Bo-Bo + 2-2 + Bo-Bo.
**Weight:** 93.00 tonnes.                  **Maximum Speed:** 80 km/h.
**Note:** For use on the Bilbao (San Nicholas)-Plentzia route.

| | | | | | | | | |
|---|---|---|---|---|---|---|---|---|
| 201/1 | 201/2 | 201/3 | 208/1 | 208/2 | 208/3 | 215/1 | 215/2 | 215/3 |
| 202/1 | 202/2 | 202/3 | 209/1 | 209/2 | 209/3 | 216/1 | 216/2 | 216/3 |
| 203/1 | 203/2 | 203/3 | 210/1 | 210/2 | 210/3 | 217/1 | 217/2 | 217/3 |
| 204/1 | 204/2 | 204/3 | 211/1 | 211/2 | 211/3 | 218/1 | 218/2 | 218/3 |
| 205/1 | 205/2 | 205/3 | 212/1 | 212/2 | 212/3 | 219/1 | 219/2 | 219/3 |
| 206/1 | 206/2 | 206/3 | 213/1 | 213/2 | 213/3 | 220/1 | 220/2 | 220/3 |
| 207/1 | 207/2 | 207/3 | 214/1 | 214/2 | 214/3 | | | |

## CLASS 300                                          2-Car Unit

**Built:** 1990-91 by CAF/AEG.
**Normal Formation:** B + B.               **Seats:** 86.
**Electrical Equipment:** AEG.             **Length:** 30.480 m.
**Motors:** 4 of 129 kW each.              **Wheel Arrangement:** Bo-Bo + 2-2.
**Weight:** 57.00 tonnes.                  **Maximum Speed:** 80 km/h.
**Note:** For use on the San Sebastian-Hendaye route.

| | | | | | | | |
|---|---|---|---|---|---|---|---|
| 301/1 | 301/3 | 304/1 | 304/3 | 307/1 | 307/3 | 310/1 | 310/3 |
| 302/1 | 302/3 | 305/1 | 305/3 | 308/1 | 308/3 | 311/1 | 311/3 |
| 303/1 | 303/3 | 306/1 | 306/3 | 309/1 | 309/3 | 312/1 | 312/3 |

## CLASS 3.100                                        3-Car Unit

**Built:** by ET/FV, Lutxana Works. Rebuilt 1988-90 by ET/FV, Atxuri Works.
**Normal Formation:** B + B + B.           **Seats:** 148.
**Electrical Equipment:**                  **Length:** 48.580 m.
**Motors:** 4 of 103 kW each.              **Wheel Arrangement:** Bo-Bo + 2-2 + 2-2.
**Weight:** 73.00 tonnes.                  **Maximum Speed:** 80 km/h.
**Note:** For use on the Bilbao (Atxuri)-San Sebastian (Amara) route.

| | | | | | | | | |
|---|---|---|---|---|---|---|---|---|
| 3.101 | 5.101 | 6.101 | 3.104 | 5.104 | 6.104 | 3.107 | 5.107 | 6.107 |
| 3.102 | 5.102 | 6.102 | 3.105 | 5.105 | 6.105 | 3.108 | 5.108 | 6.108 |
| 3.103 | 5.103 | 6.103 | 3.106 | 5.106 | 6.106 | | | |

## CLASS 3.150                                        Single Unit

**Built:** 1962-67 by ET/FV. Rebuilt 1989-90.
**Normal Formation:** B.                   **Seats:**

121

## SPAIN

**Electrical Equipment:** GEE.
**Motors:** 4 of 107 kW each.
**Weight:** 34.00 tonnes.
**Note:** For use between Eibar and Ermua.

**Length:** 17.100 m.
**Wheel Arrangement:** Bo-Bo.
**Maximum Speed:** 70 km/h.

| | | | |
|---|---|---|---|
| 3.151 | 3.152 | 3.153 | 3.154 |

## CLASS 3.500                                          3-Car Unit

**Built:** 1977-81 by CAF/AEG.
**Normal Formation:** B + B + B.
**Electrical Equipment:** AEG.
**Motors:** 4 of 121 kW each.
**Weight:** 76.00 tonnes.
**Note:** For use on the Bilbao (Atxuri)-San Sebastian (Amara) route.

**Seats:** 104.
**Length:** 47.710 m.
**Wheel Arrangement:** Bo-Bo + 2-2 + 2-2.
**Maximum Speed:** 80 km/h.

| | | | | | | | | |
|---|---|---|---|---|---|---|---|---|
| 3.501 | 5.501 | 6.501 | 3.506 | 5.506 | 6.506 | 3.511 | 5.511 | 6.511 |
| 3.502 | 5.502 | 6.502 | 3.507 | 5.507 | 6.507 | 3.512 | 5.512 | 6.512 |
| 3.503 | 5.503 | 6.503 | 3.508 | 5.508 | 6.508 | 3.513 | 5.513 | 6.513 |
| 3.504 | 5.504 | 6.504 | 3.509 | 5.509 | 6.509 | 3.514 | 5.514 | 6.514 |
| 3.505 | 5.505 | 6.505 | 3.510 | 5.510 | 6.510 | 3.515 | 5.515 | 6.515 |

# Ferrocarrils de la Generalitat de Catalunya (FGC)

**Gauges:** 1435 mm; 1000 mm.
**Electrification Systems:** 1200 V dc overhead (1435 mm); 1500 V dc overhead (1000 mm).
**Depots:**

**Route Length:** 44 km (1435 mm); 152 km (1000 mm).

## 1000 mm GAUGE ELECTRIC LOCOMOTIVES

### E1-E4                                                    C-C

**Built:** 1930 by SLM/BBC.
**Electrical Equipment:** BBC.
**Weight:** 23.00 tonnes.
**Maximum Speed:** 50 km/h (adhesion); 13 km/h (rack).
**Note:** Used on the rack operated Ribas de Freser to Nuria line.

**Continuous Rating:** 264 kW.
**Length:** 6.875 m.
**Train Supply:** Not Equipped.

| | | | |
|---|---|---|---|
| E1 | E2 | E3 | E4 |

## 1000 mm GAUGE DIESEL LOCOMOTIVES

### 700/1000 CLASS                                       Bo-Bo

**Built:** 1955-65 by Alsthom/Euskalduna/Babcock.
**Engine:** SACM MGO V12 of 590 kW at 1500 rpm.
**Weight:** 44.00 tonnes.
**Maximum Speed:** 70 km/h.

**Transmission:** Electric.
**Length:** 11.174 m.
**Train Supply:** Not equipped.

| | | | | | |
|---|---|---|---|---|---|
| 701 | 703 | 705 | 1001 | 1003 | 1022 |
| 702 | 704 | 706 | 1002 | 1009 | |

### CLASS 254                                            Co-Co

**Built:** 1990 by Meinfesa/Gm/GE.
**Engine:** GM 8-645-E3C of 1200 kW at 900 rpm.
**Transmission:** Electric.
**Weight:** 81.00 tonnes.
**Maximum Speed:** 90 km/h.

**Length:** 16.343 m.
**Train Supply:** Not equipped.

| | | |
|---|---|---|
| 254 01 | 254 02 | 254 03 |

# 1000 mm GAUGE ELECTRIC MULTIPLE UNITS

## A5-7                      2-Car Unit

**Built:** 1985-86/94 by SLM/MTM/BBC.
**Normal Formation:** B + B.            **Seats:** 112S.
**Electrical Equipment:** BBC.         **Length:** 28.080 m.
**Motors:** 2 of 361 kW each.        **Wheel Arrangement:** 1A-1A + 1A-1A.
**Weight:** 47.50 tonnes.
**Maximum Speed:** 37 km/h (adhesion); 19 km/h (rack).
**Note:** Used on the rack operated Ribas de Freser to Nuria line.

A5            A6            A7            A8

## CLASS 211                      3-Car Unit

**Built:** 1987-88 by MTM/Alsthom/Macosa.
**Normal Formation:** B + B + B.       **Seats:** 102S.
**Electrical Equipment:** Alsthom.      **Length:** 54.234 m.
**Motors:** 4 of 138 kW each.        **Wheel Arrangement:** Bo-Bo + 2-2 + 2-2.
**Weight:** 92.40 tonnes.            **Maximum Speed:** 90 km/h.

| | | | | | | | | |
|---|---|---|---|---|---|---|---|---|
| 211 01 | 281 01 | 282 01 | 211 03 | 281 03 | 282 03 | 211 05 | 281 05 | 282 05 |
| 211 02 | 281 02 | 282 02 | 211 04 | 281 04 | 282 04 | | | |

## CLASS 211                      2-Car Unit

**Built:** 1987-88 by MTM/Alsthom/Macosa.
**Normal Formation:** B + B.            **Seats:** 70S.
**Electrical Equipment:** Alsthom.      **Length:** 36.492 m.
**Motors:** 4 of 138 kW each.        **Wheel Arrangement:** Bo-Bo + 2-2.
**Weight:** 66.10 tonnes.            **Maximum Speed:** 90 km/h.

| | | | | | |
|---|---|---|---|---|---|
| 211 06 | 282 06 | 211 08 | 282 08 | 211 09 | 282 09 | 211 10 | 282 10 |
| 211 07 | 282 07 | | | | |

## CLASS 5000                    3-Car Unit

**Built:** 1960-64 by Naval. Rebuilt 1985-90 by FGC Rocafort.
**Normal Formation:** B + B + B.       **Seats:** 122S.
**Electrical Equipment:**             **Length:** 48.300 m.
**Motors:** 8 of 118 kW each.        **Wheel Arrangement:** Bo-Bo + 2-2 + Bo-Bo.
**Weight:** 92.50 tonnes.            **Maximum Speed:** 75 km/h.

| | | | | | | | | |
|---|---|---|---|---|---|---|---|---|
| 5001 | 5009 | 5103 | 5007 | 5102 | 6004 | 5008 | 5106 | 6005 |
| 5002 | 5101 | 6001 | | | | | | |

## CLASS 5000                    2-Car Unit

**Built:** 1960-64 by Naval. Rebuilt 1985-90 by FGC Rocafort.
**Normal Formation:** B + B.            **Seats:** 80S.
**Electrical Equipment:**             **Length:** 32.200 m.
**Motors:** 4 of 118 kW each.        **Wheel Arrangement:** Bo-Bo + 2-2.
**Weight:** 56.00 tonnes.            **Maximum Speed:** 75 km/h.

| | | | | | |
|---|---|---|---|---|---|
| 5003 | 6101 | 5005 | 6002 | 5006 | 6003 | 5004 | 6102 |

## CLASS 8000                    3-Car Unit

**Built:** 1949 by SIG/MFO. Acquired from Appenzeller Bahn in 1994.
**Normal Formation:**               **Seats:**
**Electrical Equipment:**             **Length:**
**Motors:**                      **Wheel Arrangement:**
**Weight:**                      **Maximum Speed:** 75 km/h.

| | | | | | |
|---|---|---|---|---|---|
| 8001 | 8101 | 8201 | 8002 | 8102 | 8202 |

# SPAIN

## 1000 mm GAUGE DIESEL MULTIPLE UNITS

## CLASS 3000 — 3-Car Unit

**Built:** 1958-68 by Ferrostahl/MAN. Rebuilt 1986-88 by Meinfesa/MAN.
**Normal Formation:** B + B.  **Seats:** 139S.
**Engines:** Two MAN D2 156 HM2US of 155 kW per power car
**Transmission:** Hydraulic. Voith Diwabus 506.
**Length:** 50.680 m.
**Weight:** 80.50 tonnes.  **Wheel Arrangement:** Bo-Bo + 2-2 + Bo-Bo.
**Maximum Speed:** 90 km/h.

| | | | | | | | | |
|---|---|---|---|---|---|---|---|---|
| 3005 | 3004 | 3006 | 3011 | 3001 | 3008 | 3015 | 3002 | 3014 |
| 3007 | 4001 | 3010 | 3013 | 3003 | 3012 | | | |

## 1435 mm GAUGE ELECTRIC MULTIPLE UNITS

## 18 — Single Unit

**Built:** 1914-24 by Brill/FFCC Catalunya. Restored for special duties.
**Normal Formation:** B.  **Seats:** 48S.
**Electrical Equipment:** GE.  **Length:** 17.300 m.
**Motors:** 4 of 77 kW each.  **Wheel Arrangement:** Bo-Bo.
**Weight:** 38.00 tonnes.  **Maximum Speed:** 80 km/h.

18

## 401-13, 501-13, 609-14, 701-08, 903-04 — Power Car

**Built:** 1945-90 by FFCC Catalunya/FGC Rocafort.
**Normal Formation:** B.  **Seats:** 64S.
**Electrical Equipment:** Cenemesa.  **Length:** 19.296 m.
**Motors:** 4 of 110 kW each.  **Wheel Arrangement:** Bo-Bo.
**Weight:** 35.50 tonnes.  **Maximum Speed:** 80 km/h.
**Note:** Work in multiple as 3 or 4-car sets with 8xx series trailers.

| | | | | | |
|---|---|---|---|---|---|
| 401 | 408 | 502 | 509 | 611 | 705 |
| 402 | 409 | 503 | 510 | 612 | 706 |
| 403 | 410 | 504 | 511 | 613 | 707 |
| 404 | 411 | 505 | 512 | 701 | 708 |
| 405 | 412 | 506 | 513 | 702 | 903 |
| 406 | 413 | 507 | 609 | 703 | 904 |
| 407 | 501 | 508 | 610 | 704 | |

## 801-14 — Trailer Car

**Built:** 1945-90 by FFCC Catalunya/FGC Rocafort.
**Normal Formation:** B.  **Seats:** 64S.
**Length:** 19.296 m.  **Wheel Arrangement:** 2-2.
**Weight:** 24.00 tonnes.  **Maximum Speed:** 80 km/h.
**Note:** Work in multiple as 3 or 4-car sets with 4xx, 5xx, 6xx, 7xx & 9xx series power cars.

| | | | | | |
|---|---|---|---|---|---|
| 801 | 804 | 807 | 809 | 811 | 813 |
| 802 | 805 | 808 | 810 | 812 | 814 |
| 803 | 806 | | | | |

## CLASS 111 — 3-Car Unit

**Built:** 1983-87 by Alsthom/MTM/Meinfesa.
**Normal Formation:** B + C + B.  **Seats:** 176.
**Electrical Equipment:** Alsthom.  **Length:** 60.744 m.
**Motors:** 4 of 275 kW each.  **Wheel Arrangement:** Bo-Bo + 2-2 + Bo-Bo.
**Weight:** 104.00 tonnes.  **Maximum Speed:** 90 km/h.

| | | | | | | | | |
|---|---|---|---|---|---|---|---|---|
| 111 01 | 181 01 | 111 02 | 111 15 | 181 08 | 111 16 | 111 29 | 181 15 | 111 30 |
| 111 03 | 181 02 | 111 04 | 111 17 | 181 09 | 111 18 | 111 31 | 181 16 | 111 32 |
| 111 05 | 181 03 | 111 06 | 111 19 | 181 10 | 111 20 | 111 33 | 181 17 | 111 34 |
| 111 07 | 181 04 | 111 08 | 111 21 | 181 11 | 111 22 | 111 35 | 181 18 | 111 36 |
| 111 09 | 181 05 | 111 10 | 111 23 | 181 12 | 111 24 | 111 37 | 181 19 | 111 38 |
| 111 11 | 181 06 | 111 12 | 111 25 | 181 13 | 111 26 | 111 39 | 181 20 | 111 40 |
| 111 13 | 181 07 | 111 14 | 111 27 | 181 14 | 111 28 | | | |

# Ferrocarrils de la Generalitat Valenciana (FGV)

**Gauge:** 1000 mm.
**Electrification System:** 600 V dc overhead.
**Depots:**

**Route Length:** 215.7 km.

## DIESEL LOCOMOTIVES

### CLASS 1000                            Bo-Bo

**Built:** 1964-67 by Alsthom/SACM.
**Engine:** SACM MGO V12 SHR of 590 kW at 1500 rpm.
**Transmission:** Electric.
**Weight:** 44.00 tonnes.                   **Length:** 11.174 m.
**Maximum Speed:** 70 km/h.        **Train Supply:** Not equipped.

1030            1032

## DIESEL MULTIPLE UNITS

### CLASS 2300                      2-Car Unit

**Built:** 1966-73 by MAN/CAF/Euskalduna/Ateinsa. Rebuilt 1984 by CAF/Babcock/Macosa.
**Normal Formation:** B + B.          **Seats:** 92S.
**Engines:** Two Pegaso 9152/2 of 155 kW each.   **Transmission:** Hydraulic. Voith Diwabus.
**Length:** 35.000 m.                 **Wheel Arrangement:** B-2 + 2-B.
**Weight:** 45.00 tonnes.              **Maximum Speed:** 80 km/h.
**Note:** Used on the Alicante-Denia route.

| | | | | | | | |
|---|---|---|---|---|---|---|---|
| 2301 | 2302 | 2305 | 2306 | 2309 | 2310 | 2313 | 2314 |
| 2303 | 2304 | 2307 | 2308 | 2311 | 2312 | 2315 | 2316 |

## ELECTRIC MULTIPLE UNITS

### CLASS 3500                      2-Car Unit

**Built:** 1954-55 by Macosa. Rebuilt 1989.
**Normal Formation:** B + B.          **Seats:** 78S
**Electrical Equipment:**            **Length:** 36.650 m.
**Motors:** 4 of 59 kW each.         **Wheel Arrangement:** Bo-Bo + 2-2.
**Weight:** 51.00 tonnes.              **Maximum Speed:** 50 km/h.
**Note:** For use on FGV Line 3.

| | | | | | | | |
|---|---|---|---|---|---|---|---|
| 3501 | 6501 | 3503 | 6503 | 3505 | 6505 | 3506 | 6506 |
| 3502 | 6502 | 3504 | 6504 | | | | |

### CLASS 3600                      3-Car Unit

**Built:** 1982 by Babcock/AEG.
**Normal Formation:** B + B + B.        **Seats:** 63S.
**Electrical Equipment:** AEG.       **Length:** 47.160 m.
**Motors:** 4 of 122 kW each.        **Wheel Arrangement:** Bo-Bo + 2-2 + 2-2.

# SPAIN

**Weight:** 75.00 tonnes.  **Maximum Speed:** 80 Km/h.
**Note:** For use on FGV Lines 1 & 2.

| | | | | | | | | |
|------|------|------|------|------|------|------|------|------|
| 3601 | 6501 | 6601 | 3605 | 6505 | 6605 | 3608 | 6508 | 6608 |
| 3602 | 6502 | 6602 | 3606 | 6506 | 6606 | 3609 | 6509 | 6609 |
| 3603 | 6503 | 6603 | 3607 | 6507 | 6607 | 3610 | 6510 | 6610 |
| 3604 | 6504 | 6604 | | | | | | |

## CLASS 3700                              2-Car Articulated Unit

**Built:** 1986-88 by CAF/Macosa/MTM/BBC.
**Normal Formation:** B + B.  **Seats:** 102S.
**Electrical Equipment:** BBC.  **Length:** 29.800 m.
**Motors:** 4 of 94 kW each.  **Wheel Arrangement:** Bo + 2 + Bo.
**Weight:** 47.00 tonnes.  **Maximum Speed:** 80 km/h.
**Note:** For use on FGV Lines 1 & 2.

| | | | | | |
|------|------|------|------|------|------|
| 3701 | 3708 | 3715 | 3722 | 3729 | 3735 |
| 3702 | 3709 | 3716 | 3723 | 3730 | 3736 |
| 3703 | 3710 | 3717 | 3724 | 3731 | 3737 |
| 3704 | 3711 | 3718 | 3725 | 3732 | 3738 |
| 3705 | 3712 | 3719 | 3726 | 3733 | 3739 |
| 3706 | 3713 | 3720 | 3727 | 3734 | 3740 |
| 3707 | 3714 | 3721 | 3728 | | |

## CLASS 3800                              2-Car Articulated Unit

**Built:** 1993-94 by CAF/GEC Alsthom.
**Normal Formation:** B + B.  **Seats:** 65S.
**Electrical Equipment:** Siemens.  **Length:** 29.700 m.
**Motors:** 4 of 103 kW each.  **Wheel Arrangement:** Bo + 2 + Bo.
**Weight:** 29.70 tonnes.  **Maximum Speed:** 65 km/h.

| | | | | | |
|------|------|------|------|------|------|
| 3801 | 3805 | 3809 | 3813 | 3816 | 3819 |
| 3802 | 3806 | 3810 | 3814 | 3817 | 3820 |
| 3803 | 3807 | 3811 | 3815 | 3818 | 3821 |
| 3804 | 3808 | 3812 | | | |

# Serveis Ferroviaris de Mallorca (SFM)

**Gauge:** 1000 mm.  **Route:** Palma-Inca (29 km).
**Depots:**

## DIESEL MULTIPLE UNITS

## CLASS 2300                                        2-Car Unit

**Built:** 1966-73 by MAN/CAF/Euskalduna/Ateinsa. Rebuilt 1991-92.
**Normal Formation:** B + B.  **Seats:** 88S.
**Engines:** Two Pegaso 9152/2 of 155 kW each.  **Transmission:** Hydraulic. Voith Diwabus.
**Length:** 35.000 m.  **Wheel Arrangement:** B-2 + 2-B.
**Weight:** 46.00 tonnes.  **Maximum Speed:** 80 km/h.

| | | | |
|------|------|------|------|
| 2314 | 2354 | 2360 | 2365 |

## CLASS 2300                                       Single Unit

**Built:** 1966-73 by MAN/CAF/Euskalduna/Ateinsa. Rebuilt 1984 by CAF/Babcock/Macosa.
**Normal Formation:** B.  **Seats:** 44S.
**Engines:** One Pegaso 9152/2 of 155 kW.  **Transmission:** Hydraulic. Voith Diwabus.
**Length:** 17.500 m.  **Wheel Arrangement:** B-2.
**Weight:** 23.20 tonnes.  **Maximum Speed:** 80 km/h.

| | |
|------|------|
| 2358 | 2364 |

## CLASS
### 2-Car Unit

**Built:** On order from CAF.
**Normal Formation:** B + B.
**Engines:** 620 kW.     **Transmission:** Hydraulic.
**Length:**
**Weight:**
**Note:** Number series not known at the time of going to press

**Seats:** 88S.

**Wheel Arrangement:**
**Maximum Speed:** 100 km/h.

xxx1          xxx2          xxx3          xxx4

## Ferrocarril de Palma al Puerto de Sóller (FPS)

**Gauge:** 914 mm.     **Route:** Palma-Puerto de Sóller (28 km).
**Depots:** Sóller.

## DIESEL LOCOMOTIVE

## D-1
### B-B

**Built:** 1968 by Autotrade.
**Engine:** Deutz BF M 716 of 370 kW at 2000 rpm.
**Transmission:** Hydraulic. Voith L420ruZ.
**Weight:** 31.00 tonnes.     **Length:** 11.380 m.
**Maximum Speed:** 60 km/h.     **Train Supply:** Not equipped.

D-1

## ELECTRIC MULTIPLE UNITS

## CLASS AAB
### Single Unit

**Built:** 1929 by Carde & Escoriaza.
**Normal Formation:** AB.
**Electrical Equipment:** Siemens.
**Motors:** 4 of 88 kW each.
**Weight:**

**Seats:** 12F, 32S.
**Length:** 14.500 m.
**Wheel Arrangement:** Bo-Bo.
**Maximum Speed:** 60 km/h.

1          2          3          4

---

# BERLIN S-BAHN & GERMAN PRIVATE RAILWAYS

**Whilst it has not proved possible to include the above in this volume, it is hoped to publish details in a later volume in the series.**

**Please watch for our advertising or ask for your name to be put on our European Mailing List.**